IBEW Aptitidue Test Prep

Disclaimer

This book is designed to provide information and guidance to individuals preparing for the International Brotherhood of Electrical Workers (IBEW) Aptitude Test. The author and publisher have made every effort to ensure the accuracy and completeness of the information contained in this book. However, they cannot guarantee that the contents of this book fully reflect the most recent version of the IBEW Aptitude Test or that the test content will remain unchanged in the future.

This book is not endorsed, affiliated with, or sponsored by the International Brotherhood of Electrical Workers (IBEW), the Electrical Training Alliance, or any other organization involved in the administration of the IBEW Aptitude Test. The IBEW and Electrical Training Alliance names and logos are the property of their respective owners.

The author and publisher disclaim any liability or responsibility for any errors, omissions, or inaccuracies in the information provided in this book. They also disclaim any liability for any damages, losses, or injuries that may arise from the use of the information contained in this book or from any reliance on its contents.

The information in this book is provided "as is" without any express or implied warranties, including but not limited to warranties of merchantability, fitness for a particular purpose, or non-infringement. In no event shall the author or publisher be liable for any direct, indirect, incidental, special, consequential, or punitive damages arising out of the use of this book or any information contained herein.

By using this book, the reader agrees to assume full responsibility for their test preparation and acknowledges that the author and publisher shall not be held responsible for any test results or consequences arising from the use of this book. The reader is encouraged to seek additional resources and professional guidance if they require further assistance in preparing for the IBEW Aptitude Test.

All rights reserved. No part of this publication may be reproduced, distributed, or transmitted in any form or by any means, including photocopying, recording, or other electronic or mechanical methods, without the prior written permission of the publisher, except in the case of brief quotations embodied in critical reviews and certain other noncommercial uses permitted by copyright law.

SECTION 1 Page 2.
SECTION 2 Page 44.

INTRODUCTION:

I'm glad to be your guide as you prepare for the IBEW aptitude test. Let's delve into the essential aspects of the test and discuss key strategies for success.

The IBEW Aptitude Test, a crucial step toward an electrical apprenticeship, assesses foundational skills in mathematics and reading comprehension. To excel, you'll need to hone your abilities in these areas. Let's break down the test components.

First, the math portion focuses on algebra and functions. Here's a brief overview of essential topics:
1. Arithmetic operations: Add, subtract, multiply, and divide numbers confidently. Practice dealing with integers, decimals, and fractions.
2. Fractions, decimals, and percentages: Convert between these forms effortlessly, and master their applications in problem-solving.
3. Ratios and proportions: Grasp these concepts, and apply them to real-world scenarios.
4. Exponents and square roots: Understand the rules and solve expressions involving them.
5. Linear equations: Solve one-variable equations, and tackle word problems with ease.
6. Quadratic equations: Factor and solve quadratics, and apply your skills to word problems.
7. Systems of linear equations: Work with multiple variables to solve complex problems.

Next, we'll examine the reading comprehension section. This area evaluates your ability to interpret written passages, which may pertain to the electrical field or general workplace situations. Improve your skills by:

1. Actively reading: Engage with the text to better understand the main ideas and supporting details.
2. Drawing conclusions: Make inferences based on the information given.
3. Identifying purpose: Discern the author's intentions or point of view.
4. Analyzing structure: Recognize how a passage is organized to convey meaning.

As you study, remember that practice makes perfect. Allocate ample time to review these subjects and complete practice problems. Consistent effort will build your confidence and knowledge. Furthermore, consider taking timed practice exams to simulate actual test conditions, as time management is crucial.

Finally, don't forget to adopt effective test-taking strategies. Familiarize yourself with techniques for managing your time, reducing anxiety, and answering multiple-choice questions efficiently.

By embracing these principles and dedicating yourself to diligent preparation, you'll be well on your way to acing the IBEW aptitude test.

Understanding the significance of the IBEW aptitude test in relation to electrical apprenticeships is crucial. Let's dive into the reasons why the test is important.

Screening tool: The IBEW aptitude test serves as a valuable method for evaluating candidates seeking apprenticeships in the electrical trade. By assessing foundational skills in math and reading comprehension, it helps identify those who possess the necessary abilities to succeed in the field.

Fairness and objectivity: The test provides a standardized measure for comparing applicants, ensuring a fair and unbiased selection process. By focusing on the essential skills required for an electrical apprenticeship, it levels the playing field for all candidates.

Skill development: The process of preparing for the IBEW aptitude test promotes the development of key skills needed in the electrical trade. As you study, you'll not only enhance your math and reading comprehension abilities but also gain insight into the demands of the profession.

Basis for advancement: Achieving a high score on the IBEW aptitude test can bolster your chances of being accepted into a reputable apprenticeship program. Once you've completed your apprenticeship, you'll be well-positioned to launch a successful career as a journeyman electrician.

Commitment to the trade: Preparing for and performing well on the IBEW aptitude test demonstrates your dedication to the electrical field. A strong test performance can signal to potential employers that you're serious about your career and possess the foundational skills needed to excel in the industry.

In summary, the IBEW aptitude test plays a vital role in determining your eligibility for an electrical apprenticeship. By performing well on the test, you'll increase your chances of being accepted into a prestigious program and set yourself on a path toward a rewarding career in the electrical trade. Therefore, investing time and effort into test preparation is crucial for your long-term success.

Having a clear understanding of the IBEW aptitude test format and structure is essential for effective preparation. Let's explore these aspects in detail.

The IBEW aptitude test is divided into two primary sections:

1. Mathematics: This section focuses on algebra and functions, and typically features multiple-choice questions. You may encounter the following topics:
   - Arithmetic operations: Addition, subtraction, multiplication, and division involving integers, decimals, and fractions.
   - Fractions, decimals, and percentages: Conversions, comparisons, and problem-solving.
   - Ratios and proportions: Applications and problem-solving in various contexts.
   - Exponents and square roots: Simplifying expressions and solving equations.
   - Linear equations: Solving for a single variable and addressing word problems.
   - Quadratic equations: Factoring, solving, and applying to word problems.
   - Systems of linear equations: Working with multiple variables and problem-solving.

2. Reading Comprehension: In this section, you'll be asked to read and interpret passages related to the electrical field or general workplace situations. Questions will likely be multiple-choice, and your abilities in the following areas will be assessed:
   - Identifying main ideas and supporting details in a passage.
   - Drawing conclusions and making inferences based on the text.
   - Determining the author's purpose or point of view.
   - Analyzing the organization and structure of a passage.

The test is timed, and you'll generally have around 2 to 2.5 hours to complete it. The number of questions may vary, but a common format consists of 33 to 50 math questions and 36 reading comprehension questions.

To excel on the IBEW aptitude test, focus on mastering the subjects within each section, as well as developing solid test-taking strategies. Practice consistently, and consider taking timed practice exams to simulate the actual test conditions. By understanding the test format and honing your skills, you'll be well-prepared to tackle the IBEW aptitude test with confidence.

Chapter 1: Test-Taking Strategies

Employing effective test-taking strategies is crucial for success on the IBEW aptitude test. Let's explore several techniques to maximize your performance.

Time management: Since the test is timed, it's essential to pace yourself. Divide the available time by the number of questions, and try to stick to that pace. Avoid spending too much time on a single question; instead, make an educated guess and move on. You can always return to it later if time permits.

Read questions carefully: Take the time to read each question thoroughly before attempting to answer. Misreading a question may lead to incorrect assumptions and mistakes. Additionally, underline keywords or important information to stay focused on the task at hand.

Eliminate wrong answers: In multiple-choice questions, eliminate options that are clearly incorrect. This process narrows down your choices, increases the odds of selecting the correct answer, and can help with educated guessing if you're unsure.

Answer all questions: There's usually no penalty for guessing on the IBEW aptitude test. Therefore, attempt to answer every question, even if you're uncertain. An educated guess is better than leaving a question unanswered.

Break down complex problems: For challenging math questions, break the problem down into smaller, more manageable steps. Solve each part methodically, and then piece together the solution.

Skim reading passages: Read the questions related to a passage first, and then skim the passage for relevant information. This technique helps you locate pertinent details more efficiently.

Stay calm and focused: Test anxiety can negatively impact performance. Maintain a positive attitude, and practice relaxation techniques such as deep breathing or visualization. Stay focused on the present and avoid dwelling on past mistakes or worrying about future questions.

Practice, practice, practice: Familiarize yourself with the test format by taking practice exams. Simulate actual test conditions by timing yourself and answering questions in a quiet environment.

By incorporating these test-taking strategies into your preparation, you'll be better equipped to tackle the IBEW aptitude test with confidence and poise. Remember, consistent practice and a positive mindset are key to achieving success.

Effective time management is a crucial skill for success on the IBEW aptitude test. Let's explore several techniques that will help you make the most of the allotted time.
1. Familiarize yourself with the test format: Knowing the structure and types of questions on the test enables you to allocate time efficiently. By understanding what to expect, you can plan your time accordingly and avoid surprises.
2. Create a time budget: Divide the total test time by the number of questions to determine how much time you can spend on each question. Stick to this pace as closely as possible, and be mindful of the time as you work through the test.
3. Prioritize questions: Begin by answering questions you feel confident about. This approach ensures you secure points for the questions you know, and can help build momentum for tackling more challenging problems.
4. Monitor your progress: Periodically check the time and assess your progress. If you're falling behind, adjust your pace accordingly to avoid running out of time.
5. Use a systematic approach: Develop a consistent method for working through problems, particularly in the math section. By employing a step-by-step approach, you can minimize errors and save time spent on reworking problems.
6. Skim and scan: In the reading comprehension section, skim passages for key information and scan for relevant details. This technique helps you quickly locate the information needed to answer questions, conserving valuable time.
7. Know when to move on: If you're stuck on a question, make an educated guess and flag it for review. You can return to the question later if time permits, but it's essential not to waste too much time on a single problem.
8. Review flagged questions: If you have time remaining, revisit questions you flagged or left unanswered. Use the remaining time to reevaluate your answers or make educated guesses on unanswered questions.

Implementing these time management techniques can greatly enhance your performance on the IBEW aptitude test. Remember, practice is key—take timed practice exams to refine your skills and become comfortable with managing time effectively.

Multiple-choice questions are a prominent component of the IBEW aptitude test. Let's delve into some practical tips for answering these questions effectively:

Read carefully: Carefully read each question and all the answer choices before making a selection. Pay close attention to keywords and details, as they may hold the key to the correct answer.

Eliminate obvious incorrect choices: Before choosing an answer, cross out options that are clearly wrong. This process narrows down your choices and increases the probability of selecting the correct answer.

Use the process of elimination: If you're uncertain about the correct answer, systematically eliminate choices that seem unlikely. By eliminating options one by one, you can narrow down your choices and make a more informed decision.

Look for clues in the question: Sometimes, the question itself may contain hints or context that can lead you to the correct answer. Be on the lookout for these subtle clues.

Stick to your first instinct: Research has shown that your first instinct is often correct. Avoid second-guessing yourself unless you're sure that your initial choice was incorrect.

Beware of extreme answers: In some cases, answer choices containing absolutes like "always" or "never" may be less likely to be correct. However, this is not a hard and fast rule, so use your judgment based on the context of the question.

Watch out for qualifiers: Qualifiers like "sometimes," "usually," or "often" may indicate a more likely correct answer, as they allow for exceptions or variability. Again, use your judgment based on the context.

Analyze answer patterns: In some cases, you might notice patterns in the answer choices. While not a foolproof method, you can sometimes use these patterns to make an educated guess when you're uncertain.

Answer all questions: There's typically no penalty for guessing on the IBEW aptitude test, so make sure to answer every question, even if you're unsure. An educated guess is better than leaving a question unanswered.

By incorporating these tips into your test-taking approach, you can enhance your performance on multiple-choice questions and increase your chances of success on the IBEW aptitude test. Remember, practice makes perfect—include multiple-choice questions in your study routine to become more adept at handling them.

Test anxiety is a common concern among test-takers, but there are ways to reduce its impact on your performance. Here are some **strategies to help you manage anxiety** and stay focused during your IBEW aptitude test:

1. Prepare well: Adequate preparation is the best way to build confidence and reduce anxiety. Study consistently, take practice tests, and familiarize yourself with the test format to ensure you're well-prepared on exam day.
2. Develop relaxation techniques: Learn and practice techniques like deep breathing, progressive muscle relaxation, or mindfulness meditation. Use these methods to calm your nerves before and during the test.
3. Get a good night's sleep: A well-rested mind is better equipped to handle stress. Make sure to get sufficient sleep leading up to the exam, particularly the night before.

4. Maintain a balanced diet: Eating well is essential for maintaining focus and energy levels. Avoid excessive caffeine or sugar, as they may increase anxiety. Opt for nutritious meals and snacks that provide sustained energy.
5. Exercise regularly: Physical activity has been shown to reduce stress and anxiety. Engage in regular exercise leading up to the test to promote mental clarity and a sense of well-being.
6. Adopt a positive mindset: Cultivate a positive attitude towards the test. Visualize yourself succeeding, and replace negative thoughts with affirmations or encouraging self-talk.
7. Take breaks during the test: If permitted, use short breaks to stretch, take deep breaths, or engage in other relaxation techniques. These brief pauses can help clear your mind and refocus your attention.
8. Stay in the present: Concentrate on the question at hand, rather than dwelling on past mistakes or worrying about future questions. This mindfulness approach helps keep anxiety in check and maximizes your focus.
9. Practice time management: Effective time management reduces anxiety by ensuring you have ample time to answer questions. Work on your pacing during practice tests to become more comfortable with the time constraints.

By incorporating these strategies into your test preparation, you can effectively manage test anxiety and enhance your performance on the IBEW aptitude test. Remember that a certain level of anxiety is normal; it's how you manage it that makes the difference. Stay confident in your abilities, and trust in your preparation.

Adopting efficient study techniques can greatly enhance your preparation for the IBEW aptitude test. Here are some effective methods to incorporate into your study routine:

Create a study plan: Outline your study goals, the topics you need to cover, and the time you have available. Break down your study sessions into manageable chunks, allocating time for each topic, and schedule regular breaks to avoid burnout.

Use active learning strategies: Instead of passively reading or watching videos, engage with the material through techniques like summarizing, teaching others, self-testing, or creating flashcards. This active involvement helps reinforce learning and improves retention.

Focus on one topic at a time: Avoid multitasking, as it can reduce focus and retention. Concentrate on mastering one subject before moving on to the next. This approach ensures a deeper understanding of each topic.

Practice with sample questions and tests: Familiarize yourself with the test format and question types by using practice tests and sample questions. Regular practice helps improve your test-taking skills, speed, and accuracy.

Review and revise regularly: Schedule regular review sessions to reinforce your learning and prevent knowledge gaps. Revisit difficult concepts or areas where you struggle to ensure a solid understanding.

Utilize mnemonic devices: Use memory aids like acronyms, rhymes, or visual associations to help remember important information or complex concepts more easily.

Leverage multiple resources: Diversify your study materials by incorporating textbooks, videos, online resources, and practice tests. This varied approach caters to different learning styles and provides a comprehensive understanding of the subject matter.

Join or create a study group: Collaborate with peers to discuss concepts, share insights, and clarify doubts. Group discussions can provide new perspectives and enhance understanding.

Optimize your study environment: Designate a quiet, comfortable, and well-lit space for studying. Eliminate distractions like mobile devices or background noise to maximize focus and productivity.

Monitor your progress: Regularly assess your performance through self-tests, quizzes, or feedback from peers or instructors. Use this feedback to identify areas that require improvement and adapt your study strategies accordingly.

By incorporating these efficient study techniques into your preparation, you'll be better equipped to tackle the IBEW aptitude test. Remember, consistency and persistence are key to successful learning. Stay committed to your study plan and trust in your abilities to achieve your goals.

Chapter 2: Mathematics Fundamentals

To prepare for the IBEW aptitude test, you should have a strong grasp of mathematics fundamentals. Here's an overview of essential math topics and concepts to help you succeed:

1. Basic arithmetic: Be comfortable with addition, subtraction, multiplication, and division of whole numbers, fractions, and decimals. Understand the order of operations (PEMDAS) for solving equations with multiple operations.
2. Ratios and proportions: Grasp the concept of ratios, which express the relationship between two numbers. Understand proportions, which are equations that show two ratios are equal.
3. Percentages: Learn to calculate percentages, increase or decrease a value by a certain percentage, and find the percentage change between two values.
4. Exponents and roots: Understand the rules for working with exponents, including multiplication, division, and power rules. Know how to find square roots and cube roots of numbers.
5. Algebra: Master basic algebraic concepts, such as solving linear equations and inequalities, simplifying expressions, and using substitution or elimination methods. Understand the properties of exponents and radicals.
6. Geometry: Review fundamental concepts like calculating the area and perimeter of various shapes, including rectangles, triangles, and circles. Understand properties of angles, parallel lines, and triangles.
7. Coordinate geometry: Learn to plot points on a coordinate plane, find the distance between two points, and determine the slope and equation of a line.
8. Word problems: Develop the ability to translate real-life scenarios into mathematical expressions and equations. Practice solving word problems that involve algebraic and geometric concepts.
9. Data interpretation: Enhance your skills in reading and interpreting graphs and charts, such as bar graphs, line graphs, and pie charts. Understand how to analyze and draw conclusions from data presented in various formats.
10. Basic statistics: Familiarize yourself with concepts like mean, median, mode, and range, as well as simple probability calculations.

To excel in the math portion of the IBEW aptitude test, make sure you practice these fundamental concepts regularly. Use a variety of practice problems, quizzes, and sample tests to gauge your understanding and build confidence. Remember to review and revise any areas where you face difficulty, and soon you'll be well-prepared for the test!

Let's dive into **basic arithmetic operations**, which include addition, subtraction, multiplication, and division. These fundamental operations will form the foundation for more complex math concepts you'll encounter in the IBEW aptitude test.

1. Addition: When you add two or more numbers, you're finding their sum. For example, 3 + 4 = 7. You can also add decimals and fractions by ensuring that you have a common denominator when working with fractions.
2. Subtraction: Subtraction involves finding the difference between two numbers. For example, 9 - 5 = 4. When subtracting decimals and fractions, ensure that the decimal points are aligned and that you're working with a common denominator for fractions.
3. Multiplication: This operation requires you to find the product of two or more numbers. For example, 4 x 3 = 12. To multiply decimals, multiply them as you would with whole numbers and then count the total number of decimal places in both factors to place the decimal point in the product. With fractions, simply multiply the numerators and denominators separately, then simplify if possible.
4. Division: Division is the process of dividing one number by another to find the quotient. For example, 12 ÷ 3 = 4. When dividing decimals, move the decimal point in the divisor to the right until it's a whole number, then move the decimal point in the dividend the same number of places. Divide as you would with whole numbers and place the decimal point in the quotient directly above its position in the dividend. For fractions, multiply the first fraction by the reciprocal of the second fraction (i.e., flip the numerator and denominator of the second fraction) and simplify if needed.

Remember the order of operations (PEMDAS/BODMAS) when working with expressions involving multiple operations: Parentheses/Brackets, Exponents/Orders, Multiplication and Division (left to right), and Addition and Subtraction (left to right). This order ensures you perform calculations correctly.

To solidify your understanding of basic arithmetic operations, practice with a variety of problems involving whole numbers, decimals, and fractions. Mix in problems that require you to use the order of operations. As you practice and refine your skills, you'll be better prepared for the IBEW aptitude test.

Let's explore **fractions, decimals, and percentages** in depth, as they're crucial concepts for the IBEW aptitude test. We'll cover each topic and provide real-world examples to help illustrate the principles.

Fractions: A fraction is a way of representing a part of a whole. It consists of a numerator (top number) and a denominator (bottom number). For instance, 3/4 represents 3 parts out of 4 equal parts. To add or subtract fractions, find a common denominator, then add or subtract the numerators accordingly. To multiply fractions, multiply the numerators and denominators separately. To divide, multiply the first fraction by the reciprocal of the second one.

Real-world example: If you have a pizza cut into 8 slices and you eat 3 slices, you've eaten 3/8 of the pizza.

Decimals: Decimals are another way to represent parts of a whole, using a base-10 system. A decimal point separates the whole number part from the fractional part. For example, 0.25 represents 25 hundredths (25/100). Adding and subtracting decimals involve aligning the decimal points and performing the operation as you would with whole numbers. For multiplication, multiply as with whole numbers and count the decimal places in both factors to place the decimal point correctly in the product. To divide decimals, adjust the divisor and dividend as described in the previous response.

Real-world example: If you have $4.25, you have 4 whole dollars and 25 cents (25 hundredths of a dollar).

Percentages: Percentages express a number as a fraction of 100. For instance, 25% means 25 parts out of 100, or 25/100, which can be simplified to 1/4 or 0.25 as a decimal. To convert a fraction to a percentage, divide the numerator by the denominator and multiply by 100. To convert a decimal to a percentage, multiply by 100 and add the percentage symbol. To convert a percentage to a fraction or decimal, reverse these processes.

Real-world example: If 30 out of 150 people attended an event, the attendance rate is (30/150) * 100 = 20%.

By understanding these fundamental concepts and practicing problems involving fractions, decimals, and percentages, you'll be well-equipped to tackle the IBEW aptitude test.

Let's dive into the concepts of **ratios and proportions**, which are important for the IBEW aptitude test. We'll cover the basics and provide a real-world example to help you grasp the principles.

1. Ratios: A ratio is a comparison of two quantities or numbers, usually expressed as "a to b" or "a:b." Ratios can be simplified by dividing both parts by their greatest common divisor. For instance, if you have 6 oranges and 9 apples, the ratio of oranges to apples is 6:9, which simplifies to 2:3 when divided by the greatest common divisor, 3.

Real-world example: If there are 10 red balls and 20 blue balls in a box, the ratio of red to blue balls is 10:20 or, when simplified, 1:2.

2. Proportions: A proportion is an equation stating that two ratios are equal. In other words, it's a relationship between two equal ratios. Proportions are written as "a/b = c/d" or "a:b :: c:d," where a and d are the extremes, and b and c are the means. To solve a proportion, you can use the cross-multiplication method, which involves multiplying the means and the extremes.

Real-world example: Suppose you know that 2 cups of flour make 12 cookies, and you want to find out how many cookies you can make with 5 cups of flour. Set up a proportion: 2/12 = 5/x. Cross-multiplying gives 2x = 60, so x = 30. You can make 30 cookies with 5 cups of flour.

By understanding the concepts of ratios and proportions and practicing related problems, you'll be well-prepared for the IBEW aptitude test. Remember to simplify ratios whenever possible and use the cross-multiplication method for solving proportions.

Alright, let's explore the concepts of **exponents and square roots**, which are vital for the IBEW aptitude test. I'll give you a clear understanding of the principles and include a real-world example for each concept.

1. Exponents: An exponent indicates the number of times a base is multiplied by itself. It's written as a small, raised number (the exponent) to the right of the base number. For example, 3^4 (read as "3 to the power of 4") means $3 \times 3 \times 3 \times 3 = 81$. Some important rules for working with exponents include:

- Any number raised to the power of 0 equals 1 (e.g., $5^0 = 1$).
- Any number raised to the power of 1 equals the number itself (e.g., $7^1 = 7$).

Real-world example: If you're calculating compound interest, you'll use exponents. If you invest $1,000 at a 5% annual interest rate for 3 years, the formula is $A = P(1 + r/n)^{nt}$, where A is the final amount, P is the principal, r is the interest rate, n is the number of times the interest is compounded per year, and t is the number of years. In this case, $A = 1000(1 + 0.05)^3 = 1000 \times 1.157625$, so you'll have $1,157.63 after 3 years.

2. Square roots: The square root of a number is the value that, when multiplied by itself, gives the original number. It's denoted by the symbol √, such as √16 = 4, because 4 × 4 = 16. A square root is the inverse operation of squaring a number. Some useful tips for square roots:
   - The square root of a perfect square (e.g., 1, 4, 9, 16) is an integer.
   - The square root of a negative number doesn't exist among real numbers.

Real-world example: Imagine you need to find the side length of a square with an area of 64 square meters. Since the area of a square is side × side, you'll need to find the square root of the area to determine the side length. The square root of 64 is 8, so each side of the square is 8 meters long.

By mastering exponents and square roots, you'll be well-prepared for the IBEW aptitude test. Remember the basic rules for exponents and practice finding square roots of various numbers to improve your skills.

Chapter 3: Algebra and Functions

Algebra and functions are essential components of the IBEW aptitude test, so let's delve into a broad overview of these topics, complete with a real-world example.

Algebra: At its core, algebra involves using symbols (often letters) to represent numbers and express relationships between them. It allows you to solve problems by manipulating these symbols following specific rules. Key algebraic concepts include:
Variables: Symbols (e.g., x, y, z) that represent unknown values or quantities.
Expressions: Combinations of numbers, variables, and operations (e.g., $2x + 3$).
Equations: Mathematical statements where two expressions are equal (e.g., $2x + 3 = 7$).
Inequalities: Mathematical statements where one expression is greater or less than another (e.g., $2x + 3 > 7$).
Functions: Functions describe relationships between two sets of numbers, where each input (independent variable) corresponds to exactly one output (dependent variable). Functions can be represented in various ways, such as equations, tables, or graphs. Key function concepts include:
Domain: The set of all possible input values for a function.
Range: The set of all possible output values for a function.
Linear functions: Functions with a constant rate of change, represented by a straight line when graphed.
Quadratic functions: Functions that involve a variable raised to the power of 2, represented by a parabolic curve when graphed.
Real-world example: Let's say you're planning a party and need to calculate the cost based on the number of guests. The venue charges a fixed fee of $200, and the catering cost is $15 per guest. This situation can be represented as a linear function: $C(x) = 15x + 200$, where $C(x)$ is the total cost, and x is the number of guests. If you have 20 guests, you can calculate the cost by substituting x with 20: $C(20) = 15(20) + 200 = \$500$.

By gaining a solid understanding of algebra and functions, you'll be well-equipped to tackle the IBEW aptitude test. Familiarize yourself with variables, expressions, equations, and inequalities, as well as functions and their various representations. Practice applying these concepts to solve real-world problems to improve your skills.

Linear equations are essential for the IBEW aptitude test, and understanding them can greatly improve your problem-solving skills. Let's dive into the topic and explore a real-world example.

A linear equation represents a straight line when graphed. It has the general form y = mx + b, where 'y' is the dependent variable, 'x' is the independent variable, 'm' is the slope (rate of change), and 'b' is the y-intercept (the point where the line crosses the y-axis).

Key concepts related to linear equations:

1. Slope (m): The steepness of the line, which represents the rate of change between the x and y values. It's calculated as the change in y divided by the change in x (rise over run).
2. Y-intercept (b): The point where the line intersects the y-axis, indicating the value of y when x is zero.
3. Slope-intercept form: The most common form of a linear equation, written as y = mx + b. This form explicitly shows the slope and y-intercept.
4. Point-slope form: Another way to express a linear equation, written as y - y1 = m(x - x1), where (x1, y1) is a known point on the line. It's useful when you know the slope and one point on the line.

Real-world example: Imagine you're a salesperson earning a base salary of $1,000 per month, plus a commission of $50 per sale. You want to calculate your total earnings based on the number of sales (x) you make.

In this case, the linear equation is y = 50x + 1000, where y represents your total earnings, and x represents the number of sales. The slope (m) is 50, indicating that your earnings increase by $50 for each sale. The y-intercept (b) is 1000, representing your base salary when you haven't made any sales.

If you make 10 sales in a month, you can calculate your total earnings by plugging in x = 10: y = 50(10) + 1000 = $1,500.

By mastering linear equations, you'll be better equipped to tackle the IBEW aptitude test. Practice working with different forms of linear equations, calculating slopes and y-intercepts, and applying these concepts to real-world scenarios.

Solving for one variable is a vital skill for the IBEW aptitude test. Let's delve into the concept and discuss a real-world example to help you understand and apply it.

When solving for one variable, your goal is to isolate that variable on one side of the equation. This process often involves several algebraic operations, such as addition, subtraction, multiplication, or division. Here are the steps to solve for one variable:

Simplify both sides of the equation, if possible. Combine like terms and perform any necessary operations.

Use inverse operations to isolate the variable. If the variable is being added or subtracted, add or subtract the opposite value to both sides. If the variable is being multiplied or divided, multiply or divide by the inverse value to both sides.

Continue using inverse operations until the variable is alone on one side of the equation. Check your solution by substituting the value back into the original equation to see if it holds true.

Real-world example: Imagine you're planning to purchase a monthly subscription for an online service. The service offers two plans: Plan A costs $40 per month, while Plan B costs $20 per month plus a $50 setup fee. You want to find out how many months it will take for both plans to cost the same.

Let x represent the number of months, and let y represent the total cost. You can set up two equations:

Plan A: $y = 40x$
Plan B: $y = 20x + 50$

To solve for x, set the two equations equal to each other: $40x = 20x + 50$.

Now, follow these steps:

Subtract 20x from both sides to combine like terms: $20x = 50$.
Divide both sides by 20 to isolate x: $x = 2.5$.
So, after 2.5 months, both plans will have the same cost. However, since you can't have half a month, it will take 3 months for both plans to cost the same.

Understanding how to solve for one variable is crucial for the IBEW aptitude test. Practice isolating variables in various types of equations and applying these techniques to real-life situations to strengthen your problem-solving abilities.

Word problems are a crucial component of the IBEW aptitude test, as they require you to apply your math skills to real-life situations. To effectively tackle word problems, follow these steps:
   1. Read the problem carefully: Ensure you fully understand the given information and what you're asked to find. It's helpful to read the problem more than once.
   2. Identify relevant information: Pick out the important details and ignore any irrelevant information. Sometimes, extra details are added to distract or confuse you.

3. Assign variables: Use variables to represent unknown values. Clearly define what each variable represents to avoid confusion.
4. Set up equations: Use the relationships between the variables to create equations. These equations will help you find the unknown values.
5. Solve the equations: Use your algebraic skills to solve the equations and find the values of the unknown variables.
6. Verify your answer: Check your solution by plugging it back into the original word problem. Ensure that your answer makes sense in the context of the problem.

Real-world example: A construction company is hired to build a fence around a rectangular yard. The yard is 3 times longer than it is wide, and the total length of the fence needed is 160 feet. What are the dimensions of the yard?

Step 1: Read the problem carefully.
Step 2: Identify relevant information:
- Yard is rectangular.
- Length is 3 times the width.
- Total fence length is 160 feet.

Step 3: Assign variables:
- Let L represent the length of the yard.
- Let W represent the width of the yard.

Step 4: Set up equations:
- Relationship between length and width: $L = 3W$.
- Fence length: $2L + 2W = 160$ (perimeter formula for a rectangle).

Step 5: Solve the equations:
- Substitute $L = 3W$ into the fence length equation: $2(3W) + 2W = 160$.
- Simplify: $6W + 2W = 160$.
- Combine like terms: $8W = 160$.
- Divide by 8: $W = 20$ feet.
- Substitute W back into $L = 3W$: $L = 3(20) = 60$ feet.

Step 6: Verify your answer: The dimensions of the yard are 60 feet by 20 feet, which makes sense in the context of the problem.

By practicing word problems and following these steps, you'll be better prepared to tackle similar problems on the IBEW aptitude test.

Here are three real-world electrical algebra word problems that you might encounter on the IBEW aptitude test:

A circuit has a resistance of 12 ohms, and the current flowing through it is 4 amperes. Calculate the voltage across the circuit using Ohm's Law ($V = IR$).
Solution:

Resistance (R) = 12 ohms
Current (I) = 4 amperes
V = IR
V = 4 amperes * 12 ohms
V = 48 volts

The voltage across the circuit is 48 volts.

An electrician needs to install three identical light bulbs in series. The total resistance of the circuit is 30 ohms. Calculate the resistance of each light bulb.
Solution:

Total resistance (R_total) = 30 ohms
Let R represent the resistance of each light bulb.
Since the light bulbs are in series, the total resistance is the sum of the resistances of each light bulb: R_total = R + R + R

30 ohms = 3R
R = 30 ohms / 3
R = 10 ohms

The resistance of each light bulb is 10 ohms.

A transformer needs to step down the voltage from 480 volts to 120 volts. The primary coil has 1,200 turns. Calculate the number of turns required for the secondary coil.
Solution:

Primary voltage (V1) = 480 volts
Secondary voltage (V2) = 120 volts
Primary coil turns (N1) = 1,200 turns
Let N2 represent the number of turns in the secondary coil.
Using the transformer formula, V1/V2 = N1/N2:

480 volts / 120 volts = 1,200 turns / N2
4 = 1,200 turns / N2
N2 = 1,200 turns / 4
N2 = 300 turns

The secondary coil requires 300 turns.

Practicing these types of electrical algebra word problems will help you better prepare for the IBEW aptitude test.

Quadratic equations are a type of polynomial equation with a degree of 2, meaning the highest exponent of the variable is 2. They are often represented in the standard form: $ax^2 + bx + c = 0$, where a, b, and c are constants, and x is the variable. Quadratic equations play a significant role in various mathematical problems and real-world applications.

There are three primary methods to solve quadratic equations:

Factoring: Try to rewrite the equation in the form of two binomials multiplied together. Then, set each binomial equal to zero and solve for x.
Quadratic Formula: The quadratic formula, $x = (-b \pm \sqrt{b^2 - 4ac}) / 2a$, is a general method for finding the roots of a quadratic equation.
Completing the square: In this method, you rewrite the equation in the form of a perfect square trinomial, and then solve for x by taking the square root of both sides.
Real-world example: Suppose you're an electrician, and you need to determine the maximum height a projectile launched from the ground will reach if it follows a parabolic path. The height (h) of the projectile at any given time (t) can be modeled by the equation $h(t) = -16t^2 + 64t$.

This is a quadratic equation in the form of $at^2 + bt + c$ with a = -16, b = 64, and c = 0. To find the maximum height, we can use the vertex formula: t = -b / 2a. Substituting the values, we get t = -64 / (2 * -16) = 2 seconds. Now, plug t back into the equation to find the maximum height: $h(2) = -16(2)^2 + 64(2) = 64$ feet. The projectile will reach its maximum height of 64 feet after 2 seconds.

Factoring is the process of breaking down an expression into its constituent parts, typically by identifying common factors or patterns. This helps simplify expressions and solve equations.
There are several common factoring techniques you'll want to become familiar with. Let's examine each one with examples to help solidify your understanding.
1. Greatest Common Factor (GCF): This is the largest number or term that divides two or more expressions evenly. To factor using the GCF, first find the GCF of the terms and then divide each term by the GCF.

Example: Factor $12x^2 + 16x$. GCF of $12x^2$ and 16x is 4x. Factored form: $4x(3x + 4)$.

2. **Difference of Squares:** This technique applies to expressions in the form of $a^2 - b^2$, which can be factored into $(a + b)(a - b)$.

Example: Factor $x^2 - 25$. Here, a = x and b = 5. Factored form: $(x + 5)(x - 5)$.

3. **Trinomial Factoring:** This involves factoring quadratic expressions in the form of $ax^2 + bx + c$. For simple cases where a = 1, look for two numbers that multiply to c and add to b.

Example: Factor $x^2 + 5x + 6$. Find two numbers that multiply to 6 and add to 5: 2 and 3. Factored form: $(x + 2)(x + 3)$.

4. **Factoring by Grouping:** This method is useful when you have four or more terms. Group the terms into pairs and factor out the GCF from each pair. If done correctly, you'll have a common binomial factor.

Example: Factor $x^3 - x^2 + 3x - 3$. Group terms: $(x^3 - x^2) + (3x - 3)$. Factor GCF from each group: $x^2(x - 1) + 3(x - 1)$. Factor out the common binomial: $(x - 1)(x^2 + 3)$.

Remember to practice factoring problems with a variety of expressions to build your skills and become proficient in this area.

here are a couple of real-world electrician-related word problems involving quadratic equations:

Power Loss in a Cable:

An electrician installs a 100-meter long underground cable between two electrical panels. Due to the resistance in the cable, there is some power loss. The power loss (P) in watts can be modeled by the equation $P = R * I^2$, where R is the resistance in ohms and I is the current in amperes. The resistance R of the cable is given by $R = k * L / A$, where k is the resistivity, L is the cable length, and A is the cross-sectional area. The electrician wants to limit the power loss to 50 watts with a maximum current of 10 amperes. If the resistivity (k) of the cable material is 0.0001 ohm-meters, what is the minimum cross-sectional area of the cable?

First, we know that P = 50 watts and I = 10 amperes. So, $R = P / I^2 = 50 / (10^2) = 0.5$ ohms.

Now, we have the equation 0.5 = (0.0001)(100) / A, which simplifies to $A^2 - 20A + 0 = 0$. This is a quadratic equation. Solving it, we find that A ≈ 0.245 square meters. So, the electrician should use a cable with a cross-sectional area of at least 0.245 square meters.

Maximum Height of a Projectile:

An electrician is using a wire puller to send a cable over a tall obstacle. The height (h) in meters of the projectile (the cable end with a weight) can be modeled by the equation $h = -4.9t^2 + vt + s$, where t is the time in seconds, v is the initial upward velocity in meters per second, and s is the initial height in meters. The electrician releases the projectile with an initial velocity of 20 m/s from a height of 2 meters. What is the maximum height the projectile will reach above the ground?

The height of the projectile is given by $h(t) = -4.9t^2 + 20t + 2$.

To find the maximum height, we need to find the vertex of this quadratic equation. The vertex's x-coordinate (time) can be found using the formula $t = -b / 2a$, where $a = -4.9$ and $b = 20$. This gives $t \approx 2.04$ seconds.

Now, we can find the maximum height by plugging this time into the height equation: $h(2.04) \approx -4.9(2.04)^2 + 20(2.04) + 2 \approx 22.2$ meters. So, the projectile will reach a maximum height of about 22.2 meters above the ground.

These examples demonstrate how quadratic equations can be applied to real-world electrician problems. Keep practicing to get comfortable with solving word problems in various contexts.

Systems of linear equations involve multiple linear equations with the same variables. The goal is to find the values of the variables that satisfy all the equations simultaneously.

There are three primary methods to solve systems of linear equations: graphing, substitution, and elimination. We'll explore each method with examples to solidify your understanding.

1. Graphing: In this method, you graph each equation on the same coordinate plane and look for the point(s) where the lines intersect.

Example: Solve the system of equations: $y = 2x + 1$ $y = -x + 3$
Graph both lines and find the intersection point. In this case, the intersection is at the point (1, 3). So, the solution is $x = 1$ and $y = 3$.

2. Substitution: This technique involves solving one equation for one variable and then substituting that expression into the other equation(s) to eliminate the variable.

Example: Solve the system of equations: $y = 2x + 1$ $y = -x + 3$
Since both equations are already solved for y, substitute the expression for y from the first equation into the second: $2x + 1 = -x + 3$. Solve for x: $3x = 2$, so $x = 2/3$. Now, substitute this value back into either equation to find y: $y = 2(2/3) + 1$, which gives $y = 7/3$. The solution is $x = 2/3$ and $y = 7/3$.

3. Elimination: In this method, you manipulate the equations to eliminate one variable by adding or subtracting the equations.

Example: Solve the system of equations: $3x - 2y = 7$ $2x + y = 1$

First, multiply the second equation by 2 to make the coefficients of y equal: $4x + 2y = 2$. Now, add the two equations to eliminate the y variable: $7x = 9$. Divide by 7 to find x: $x = 9/7$. Finally, substitute this value back into either equation to find y. Using the second equation: $2(9/7) + y = 1$, we find that $y = -5/7$. The solution is $x = 9/7$ and $y = -5/7$.

As you prepare for the IBEW aptitude test, practice solving systems of linear equations using these three methods. Familiarize yourself with different types of linear systems, including those with one solution, no solution, or infinitely many solutions. Developing a strong foundation in this area will serve you well on the test.

Solving for multiple variables is a valuable skill for your IBEW aptitude test. It involves finding the values of variables that satisfy a set of equations. To tackle this topic, let's examine a few methods you can use to solve for multiple variables, focusing on systems of linear equations.

Substitution: This technique involves solving one equation for a variable and then substituting that expression into the other equation(s) to eliminate the variable.
Example: Solve the system of equations:
$y = 3x - 4$
$x + 2y = 6$

First, solve the first equation for y: $y = 3x - 4$. Next, substitute this expression into the second equation: $x + 2(3x - 4) = 6$. Simplify and solve for x: $7x = 14$, so $x = 2$. Now, substitute x back into the first equation to find y: $y = 3(2) - 4$, which gives $y = 2$. The solution is $x = 2$ and $y = 2$.

Elimination: This method involves manipulating the equations to eliminate one variable by adding or subtracting the equations.
Example: Solve the system of equations:
$2x + y = 5$
$3x - 2y = 1$

First, multiply the first equation by 2 to make the coefficients of y equal: $4x + 2y = 10$. Now, add the two equations to eliminate the y variable: $7x = 11$. Divide by 7 to find x: $x =$

11/7. Finally, substitute this value back into either equation to find y. Using the first equation: 2(11/7) + y = 5, we find that y = -3/7. The solution is x = 11/7 and y = -3/7.

Matrix Method: This technique uses matrices to represent the system of linear equations and employs matrix operations to solve for the variables.
Example: Solve the system of equations:
2x + y = 5
3x - 2y = 1

Represent the system as an augmented matrix:
[2 1 | 5]
[3 -2 | 1]

Use row operations to transform the matrix into reduced row-echelon form:
[1 0 | 11/7]
[0 1 | -3/7]

The solution is x = 11/7 and y = -3/7.

As you prepare for the IBEW aptitude test, practice solving for multiple variables using these methods. Be sure to work with different types of systems, including those with one solution, no solution, or infinitely many solutions. Strengthening your skills in solving for multiple variables will be immensely helpful on the test.

Let's explore a couple of real-world electrician-related word problems involving systems of linear equations.

1. Circuit Analysis: An electrician is analyzing a circuit with two parallel resistors, R1 and R2, connected to a voltage source. The total resistance (Rt) in the circuit is given by the formula 1/Rt = 1/R1 + 1/R2. The electrician measures the total current (It) in the circuit as 15 A. Ohm's Law states that voltage (V) equals current (I) times resistance (R), or V = IR. The electrician finds that the voltage drop across R1 is 45 V. Calculate the values of R1 and R2.

First, let's write down the equations we need: V1 = R1 * I1 V2 = R2 * I2 V1 + V2 = 45 V I1 + I2 = 15 A

Now we have a system of four equations with four variables (V1, V2, I1, and I2). Solve the equations to find the values of R1 and R2. The solution is R1 ≈ 5 ohms and R2 ≈ 9 ohms.

2. Wire Length Calculation: An electrician is installing two types of wires, A and B, in a building. Each wire has a different cost per foot: Wire A costs $2 per foot, and Wire B costs $4 per foot. The electrician has a budget of $800 for this project. Additionally, the electrician needs to maintain a specific length ratio of 3:5 between Wire A and Wire B to meet the project requirements. Determine how many feet of each wire type the electrician should purchase.

Let x be the length of Wire A and y be the length of Wire B. We have two equations:
$2x + 4y = 800$ (the cost constraint) $x / y = 3 / 5$ (the length ratio constraint)
Now we have a system of two linear equations with two variables, x and y. Solve the equations to find the lengths of Wire A and Wire B. The solution is x = 240 feet and y = 400 feet.

Practicing these real-world electrician-related word problems will help you become more comfortable with solving systems of linear equations in various contexts, an essential skill for your IBEW aptitude test.

Chapter 4: Geometry

Geometry is an essential topic for your IBEW aptitude test, so let's dive deep and explore its key concepts. We'll cover some fundamental principles and provide real-world examples to illustrate the ideas.

Points, Lines, and Planes:
In geometry, points represent specific locations, lines are one-dimensional figures that extend infinitely in both directions, and planes are two-dimensional flat surfaces that extend infinitely. Understanding these basic elements is crucial for solving geometric problems.

Angles:
An angle is formed by two rays that share a common endpoint (vertex). Angles are measured in degrees. There are several types of angles: acute (less than 90 degrees), right (90 degrees), obtuse (between 90 and 180 degrees), and straight (180 degrees).
Example: Electricians often encounter right angles when installing electrical components in walls or ceilings, as these structures typically meet at 90-degree angles.

Triangles:
Triangles are three-sided polygons. They can be classified by their angles (acute, right, obtuse) or by their sides (equilateral, isosceles, scalene). The sum of the angles in a triangle is always 180 degrees.
Example: An electrician might use the Pythagorean theorem ($a^2 + b^2 = c^2$) to calculate the length of a wire needed to connect two points in a right-angled triangle.

Quadrilaterals:
Quadrilaterals are four-sided polygons. Some common types include squares, rectangles, parallelograms, trapezoids, and rhombuses. Each type has specific properties that can help solve geometric problems.
Example: When installing solar panels on a rectangular roof, an electrician might need to calculate the area of the rectangle to determine the number of panels that can fit.

Circles:
A circle is a set of points equidistant from a center point. The distance from the center to any point on the circle is called the radius. The diameter is twice the radius. Circumference is the distance around the circle, and area is the space enclosed by the circle.

Example: An electrician might use the area of a circle to find the space needed to install a round electrical junction box.

Perimeter and Area:
Perimeter is the distance around a shape, and area is the space enclosed by the shape. Different formulas are used to calculate the perimeter and area of various shapes.
Example: To determine the amount of wire conduit needed for a room, an electrician might calculate the perimeter of the room.

Practice working with these geometric concepts and apply them to real-world electrician-related problems. Familiarity with these principles will serve you well on your IBEW aptitude test.

Let's explore **basic geometry concepts and terminology** to help you prepare for your IBEW aptitude test.

1. Point: A point represents a location in space. It has no dimensions (length, width, or height). In geometry, points are often labeled with capital letters, such as A, B, and C.
2. Line: A line is a straight, one-dimensional figure that extends infinitely in both directions. Lines are often labeled with lowercase letters or by using two points on the line, such as line l or line AB.
3. Line segment: A line segment is a part of a line with two endpoints. It's represented by the two endpoints and a line over them, like AB.
4. Ray: A ray is a part of a line with a single endpoint, called the initial point, and extends infinitely in the other direction. It's represented by the initial point and another point on the ray, such as ray BA.
5. Plane: A plane is a two-dimensional flat surface that extends infinitely in all directions. Planes are often labeled with capital italic letters, such as plane M.
6. Collinear points: Points that lie on the same line are called collinear points.
7. Coplanar points: Points that lie on the same plane are called coplanar points.
8. Angle: An angle is formed by two rays that share a common endpoint (vertex). Angles are measured in degrees and labeled with a lowercase Greek letter theta (θ) or by the vertex and points on each ray, such as $\angle ABC$.
9. Parallel lines: Two lines are parallel if they are coplanar and never intersect.
10. Perpendicular lines: Two lines are perpendicular if they intersect at a 90-degree angle.
11. Congruent figures: Two geometric figures are congruent if they have the same shape and size.

12. Similar figures: Two geometric figures are similar if they have the same shape but not necessarily the same size.

Real-world example: Electricians often work with perpendicular lines when installing electrical conduits along walls and ceilings, as these structures typically meet at right angles.

By understanding these basic geometry concepts and terminology, you'll be better equipped to tackle geometry problems on the IBEW aptitude test. Practice using these terms and concepts in various geometric contexts to solidify your understanding.

Understanding the **perimeter, area, and volume** of common shapes is essential for your IBEW aptitude test preparation. We'll maintain a human-like tone and provide real-world examples where applicable.

Perimeter:
Perimeter refers to the total distance around a two-dimensional shape. Here are the formulas for the perimeter of common shapes:
Rectangle: $P = 2l + 2w$ (l = length, w = width)
Square: $P = 4s$ (s = side length)
Triangle: $P = a + b + c$ (a, b, and c are the side lengths)

Area:
Area is the amount of space enclosed by a two-dimensional shape. Here are the formulas for the area of common shapes:
Rectangle: $A = lw$ (l = length, w = width)
Square: $A = s^2$ (s = side length)
Triangle: $A = 0.5bh$ (b = base, h = height)
Circle: $A = \pi r^2$ (r = radius, $\pi \approx 3.14159$)
Real-world example: An electrician may need to calculate the area of a rectangular wall to determine how many electrical outlets are required per code.

Volume:
Volume is the amount of space occupied by a three-dimensional object. Here are the formulas for the volume of common shapes:
Rectangular prism: $V = lwh$ (l = length, w = width, h = height)
Cube: $V = s^3$ (s = side length)
Cylinder: $V = \pi r^2 h$ (r = radius, h = height, $\pi \approx 3.14159$)
Sphere: $V = 4/3 \pi r^3$ (r = radius, $\pi \approx 3.14159$)
Cone: $V = 1/3 \pi r^2 h$ (r = radius, h = height, $\pi \approx 3.14159$)

Real-world example: An electrician might need to calculate the volume of a cylindrical conduit to determine its capacity for holding wires.

By mastering these formulas for perimeter, area, and volume, you'll be better prepared for your IBEW aptitude test. Practice applying these formulas to various problems and real-world scenarios to strengthen your understanding.

Here are three real-world geometry word problems related to electrician work, with a human-like tone and friendly, instructive approach.
1. An electrician needs to install a rectangular solar panel on a rooftop. The solar panel is 4 feet wide and 6 feet long. To abide by the building code, there must be a 2-foot clearance around the entire panel. What is the total area occupied by the solar panel and the required clearance?
2. While wiring a new home, an electrician needs to run a wire diagonally across a rectangular room that measures 12 feet by 16 feet. What length of wire is needed to span the diagonal of the room? (Hint: Use the Pythagorean theorem to solve this problem.)
3. An electrician is installing cylindrical recessed lights in a ceiling. Each light has a diameter of 6 inches and a depth of 4 inches. Calculate the volume of the cylindrical space the lights will occupy in the ceiling. How many cubic inches of space will be used by three recessed lights?

Chapter 5: Reading Comprehension Strategies

Reading comprehension is an essential skill for the IBEW aptitude test.

Preview the text: Before diving into a passage, take a moment to skim the headings, subheadings, and first sentences of each paragraph. This will help you get a sense of the structure and main ideas, allowing you to read more efficiently.

Read actively: Engage with the text by asking questions, making predictions, and visualizing the content as you read. This active approach will help you stay focused and retain information more effectively.

Take notes: Jot down key points, ideas, or questions as you read. This will not only help you remember important information but also make it easier to review later.

Summarize: After reading a section or the entire passage, take a moment to summarize what you've learned in your own words. This will reinforce your understanding and help you retain the information.

Make connections: Relate the information in the passage to your own experiences or prior knowledge. This will help deepen your understanding and make the content more memorable.

Monitor your comprehension: While reading, periodically check if you're understanding the material. If you're struggling with a particular section, try re-reading it or breaking it down into smaller parts.

Use context clues: If you encounter an unfamiliar word, try to deduce its meaning from the surrounding words and sentences. This will help you maintain your flow and build your vocabulary.

Reflect and review: After finishing the passage, reflect on what you've learned and review your notes. This will solidify your understanding and help you remember the material more effectively.

Real-world example: Electricians may need to read and comprehend technical manuals or installation guides. Applying these strategies will help them understand complex information and perform their job more efficiently.

Active reading techniques are crucial for improving comprehension, especially when preparing for the IBEW aptitude test.

1. Previewing: Before reading, skim the text to get an idea of its structure and main points. Look at headings, subheadings, and the first sentences of each

paragraph. This helps you create a mental roadmap and improves your understanding as you read.
2. Questioning: As you read, ask questions like "What is the author's purpose?" or "What do I expect to learn from this section?" This helps you stay engaged and focused on the content.
3. Annotating: Make notes or highlight key points in the text. Write down questions or connections you make while reading. This keeps you actively involved with the material and aids in retention.
4. Visualizing: Create mental images of the information as you read. This can help you better understand complex concepts and make the content more memorable.
5. Outlining: As you read, create an outline that captures the main ideas and supporting details. This helps you organize the material, making it easier to review and remember.
6. Summarizing: After finishing a section, write a brief summary of the main ideas in your own words. This reinforces your understanding and retention of the material.
7. Reflecting: Consider how the information in the text relates to your own experiences or prior knowledge. Making connections helps you better understand and remember the content.

Real-world example: Imagine you're an electrician reading a technical manual for a new piece of equipment. Using active reading techniques, you can efficiently comprehend the material and apply the information to your work.

Drawing conclusions and making inferences are critical skills for reading comprehension, especially when preparing for the IBEW aptitude test.

Drawing conclusions: When you draw a conclusion, you synthesize information from the text to arrive at an overall understanding or judgment. To do this effectively:
1. Identify the main ideas: As you read, take note of the most important points in the text.
2. Look for supporting details: These could be examples, explanations, or evidence that reinforce the main ideas.
3. Consider the author's purpose: Think about why the author wrote the text and what they're trying to convey.
4. Use your background knowledge: Connect the information in the text to your own experiences or prior knowledge.

5. Synthesize the information: Combine your understanding of the main ideas, supporting details, author's purpose, and background knowledge to form a well-rounded conclusion.

Making inferences: Inferences involve using clues in the text, along with your background knowledge, to read between the lines and understand implied meanings. To make inferences effectively:
1. Observe textual clues: Pay attention to words, phrases, or passages that hint at the author's intended meaning.
2. Use context: Consider the surrounding sentences or paragraphs to help interpret the meaning of a specific word or phrase.
3. Apply your background knowledge: Draw on your own experiences or previous learning to make connections and fill in gaps.
4. Ask questions: As you read, consider what might be implied or suggested by the text.
5. Evaluate your inferences: Assess whether your inferences are logical and supported by the text and context.

Real-world example: As an electrician, you might read a safety manual that mentions the importance of using insulated tools. By drawing conclusions, you can understand that using insulated tools reduces the risk of electric shock. By making inferences, you can deduce that using non-insulated tools in specific situations might lead to dangerous consequences.

Understanding the author's purpose and point of view is crucial for reading comprehension, especially when preparing for the IBEW aptitude test.

Author's purpose:
The author's purpose is the reason they have written a text. It's important to identify this to better understand the main ideas and context. There are typically three main purposes:

Inform: The author aims to provide information, facts, or explanations about a topic.
Persuade: The author tries to convince the reader to accept a specific viewpoint or take a certain action.
Entertain: The author intends to engage the reader with a story or interesting content.
To identify the author's purpose, look for clues like the text's structure, tone, and the type of information provided. Consider the intended audience and how the author addresses them.

Author's point of view:

The point of view refers to the perspective or stance from which the author writes. This influences the way information is presented, the tone, and the opinions expressed. There are three main points of view:

First person: The author writes from their own perspective, using words like "I" and "we."
Second person: The author addresses the reader directly, using the word "you."
Third person: The author writes from an outside perspective, using words like "he," "she," and "they."
To understand the author's point of view, pay attention to the pronouns used and the way the author presents information. Look for any bias or personal opinions that may be influencing the content.

Real-world example: As an electrician, you might read an article about new safety regulations in your field. The author's purpose could be to inform you about these changes, and their point of view might be from an industry expert perspective. Recognizing this can help you understand the context of the article and the author's credibility.

By comprehending the author's purpose and point of view, you'll gain a deeper understanding of the text, which will enhance your reading comprehension skills for the IBEW aptitude test. Keep practicing these techniques with various texts to solidify your abilities.

Analyzing the organization and structure of a passage is vital for proper reading comprehension, and it can be especially helpful when preparing for the IBEW aptitude test.

Organization and structure refer to the way an author arranges ideas and information in a passage. By examining these aspects, you can better understand the main ideas, supporting details, and the relationships between them. Here are some common structures you might encounter:

Chronological order: Events or information are presented in the order they occurred. This structure is common in historical accounts or step-by-step instructions.
Problem and solution: The author presents a problem and then explains one or more solutions to address it.
Cause and effect: The author discusses an event (cause) and its consequences (effects).

Compare and contrast: The author explores similarities and differences between two or more subjects.

Descriptive: The author provides a detailed description of a person, place, or thing.

To analyze the organization and structure of a passage, follow these steps:

Read the passage carefully and identify the main idea or topic.

Look for transition words or phrases that indicate a change in focus, such as "however," "on the other hand," "for example," or "as a result."

Determine the relationships between ideas, events, or concepts in the passage.

Identify the structure the author uses to organize the information.

Consider how the structure and organization contribute to the overall meaning and effectiveness of the passage.

Real-world example: As an electrician, you might read a technical manual explaining how to install a specific electrical component. The manual likely uses a chronological order structure to present the installation steps in a logical sequence. Understanding this organization can help you follow the instructions accurately and safely.

Chapter 6: Vocabulary and Context Clues

Developing a strong vocabulary and the ability to use context clues is essential for the reading comprehension section of the IBEW aptitude test.

Vocabulary refers to the collection of words you know and understand. A broad vocabulary allows you to comprehend a wider range of texts, including those that contain technical terms or industry-specific language.

Context clues are hints or cues within a passage that help you determine the meaning of unfamiliar words or phrases. There are several types of context clues you can look for:

1. Definition: The meaning of the word is provided directly in the sentence or nearby. Example: "The oscilloscope, a device that measures electronic signals, displayed the waveform."
2. Synonym: A word with a similar meaning is used in the same sentence or nearby. Example: "The electrician was diligent and meticulous when installing the new wiring."
3. Antonym: A word with an opposite meaning is used in the same sentence or nearby. Example: "Unlike the experienced electrician, the apprentice was unfamiliar with the equipment."
4. Example: An example is provided to help illustrate the meaning of the word. Example: "The electrician used various tools, such as pliers, screwdrivers, and wire strippers."
5. Inference: The overall context of the passage provides clues to the word's meaning. Example: "The electrician meticulously inspected the wiring for any faults or issues."

To improve your vocabulary and context clue skills, follow these steps:

1. Read a variety of texts related to electrical work and general topics to encounter new words and phrases.
2. When you come across an unfamiliar word, try to determine its meaning using context clues before looking it up in a dictionary.
3. Keep a list of new words you learn, and review them regularly to reinforce your understanding.
4. Practice using the new words in context, either in conversation or writing, to solidify your knowledge.

By working on your vocabulary and context clue abilities, you'll enhance your reading comprehension skills and be better prepared for the IBEW aptitude test.

Expanding your vocabulary is an essential part of preparing for the IBEW aptitude test, as well as any endeavor that involves reading comprehension or communication. An extensive vocabulary will not only improve your test-taking abilities but also enhance your overall understanding and expression of ideas. To expand your vocabulary, consider the following strategies:

1. Read widely: A surefire way to expand your vocabulary is to read a diverse array of materials. Books, newspapers, magazines, and online articles can all expose you to new words, phrases, and concepts. As you encounter unfamiliar terms, make an effort to understand their meanings in context and look them up if necessary.
2. Keep a vocabulary journal: Whenever you come across a new word, jot it down in a dedicated journal or notebook. Record the definition, the context in which you found the word, and any synonyms or antonyms. Reviewing these entries regularly will help reinforce your learning.
3. Learn root words, prefixes, and suffixes: Understanding the building blocks of words can enable you to deduce the meanings of unfamiliar terms. For example, the prefix "dis-" often indicates negation or reversal, as in "disagree" or "disconnect." Recognizing these patterns can help you quickly decipher the meanings of new words.
4. Use flashcards: Create flashcards with a word on one side and its definition, along with an example sentence, on the other. Regularly reviewing these flashcards can help you commit new vocabulary to memory. You can also use digital flashcard apps for added convenience.
5. Engage in word games and puzzles: Crossword puzzles, word searches, and games like Scrabble or Boggle can be fun ways to reinforce your vocabulary while exercising your brain. These activities challenge you to recall and apply the words you've learned.
6. Practice using new words: Actively incorporating newly learned words into your speech and writing will help solidify your understanding of them. When you encounter a new word, try to use it in a sentence or two over the next few days to reinforce its meaning.
7. Join a discussion group or book club: Engaging in conversations or discussions about various topics will expose you to new words and ideas. Additionally, discussing what you've read with others can help you better understand and remember the material.

By employing these strategies consistently, you'll expand your vocabulary and become better prepared for the IBEW aptitude test, as well as enhance your overall communication skills.

Context clues are essential for determining the meanings of unfamiliar words when you're reading a passage. Using context clues allows you to make educated guesses about the meaning of a word without needing to consult a dictionary. Here are some strategies for using context clues effectively:

Look for synonyms and antonyms: Often, authors will include words or phrases with similar or opposite meanings close to the unfamiliar word. These can help you understand the general sense of the word. For example, if you encounter the word "tenacious" in a sentence like "She was a tenacious and determined athlete, never giving up despite the challenges," you can deduce that "tenacious" means persistent or unyielding.

Pay attention to examples and explanations: Sometimes, an author will provide examples or explanations to illustrate the meaning of a new word. For instance, if you read, "The view from the summit was breathtaking, with an expansive panorama of mountains, forests, and lakes," you can infer that "panorama" refers to a wide, sweeping view.

Examine the surrounding words: The words immediately before or after the unfamiliar word can provide clues to its meaning. Look for prepositions, conjunctions, or punctuation marks that might link the unknown word to a related idea or concept.

Consider the sentence structure: The way a sentence is structured can help reveal the function and meaning of a word. For example, if the unfamiliar word is the subject or object of a verb, it's likely a noun. If it's modifying a noun or pronoun, it's probably an adjective or adverb.

Evaluate the tone and mood of the passage: The overall tone and mood of the text can offer hints about the meaning of a word. If the passage is positive or upbeat, the word may have a positive connotation. If the passage is negative or critical, the word might have a negative connotation.

Break down the word into smaller parts: Analyzing the prefixes, suffixes, and root words can give you clues about the word's meaning. For example, if you see the word "misconstrue," you can break it down into "mis-" (wrongly) and "construe" (interpret), giving you the idea that it means to interpret something incorrectly.

By practicing these strategies, you'll become more adept at using context clues to decipher unfamiliar words. This skill will be invaluable for the IBEW aptitude test and any reading-intensive tasks you encounter in the future.

Practicing reading passages is a valuable way to prepare for the IBEW aptitude test. By working through different passages, you can improve your reading comprehension skills and get a feel for the types of questions you may encounter on the test. Here are some tips and strategies to make the most of your practice:

1. Select diverse passages: Choose practice passages from various sources and genres, such as technical articles, news reports, opinion pieces, and literary excerpts. This will help you become comfortable with different writing styles and topics.
2. Time yourself: Reading comprehension on the IBEW aptitude test is time-limited. Simulate the test conditions by setting a timer and practicing your ability to read and answer questions quickly and accurately.
3. Apply active reading techniques: As you read, focus on understanding the main idea, key points, and supporting details. Make notes or annotations as you go, especially for complex or dense sections.
4. Answer questions strategically: After reading a passage, try to answer the questions without referring back to the text. This will help you gauge how well you've understood the material. If you're unsure about a question, use the process of elimination to narrow down your options.
5. Review your answers: After completing a practice passage, go back and review your answers. Identify any incorrect responses and analyze why you chose the wrong answer. This will help you recognize patterns in your mistakes and improve your overall comprehension.
6. Reflect on your progress: Regularly assess your performance to identify areas that need improvement. Are there certain types of questions or topics that consistently trip you up? Focus on these areas and adjust your strategies accordingly.
7. Learn from real-world examples: Apply your reading comprehension skills to everyday situations. Read articles, books, or documents related to your field, such as electrical engineering or construction. Analyzing real-world materials will not only expand your knowledge but also help you become a more efficient reader.

By following these tips and consistently practicing with a variety of reading passages, you'll be well-prepared for the reading comprehension section of the IBEW aptitude test.

Electrical industry-related passages on the IBEW aptitude test are designed to assess your comprehension of materials that are relevant to the field. To prepare for these passages, familiarize yourself with essential topics and concepts that you may encounter. Here are some key areas to focus on, along with strategies for understanding them:

1. Basic electrical theory: Understand fundamental concepts, such as voltage, current, resistance, power, and energy. Familiarize yourself with Ohm's Law, Kirchhoff's Laws, and the relationships between these quantities. "

2. Electrical components and devices: Learn about common electrical components, such as resistors, capacitors, inductors, switches, and transformers. Understand their functions, symbols, and how they work in electrical circuits.
3. Circuit analysis: Study different circuit analysis techniques, such as series and parallel circuits, Thevenin's theorem, and Norton's theorem. Practice solving problems that involve calculating voltages, currents, and resistances in various configurations.
4. Electrical safety: Acquaint yourself with safety guidelines, such as lockout/tagout procedures, personal protective equipment, and grounding methods. Understand the importance of following safety protocols and the consequences of not adhering to them.
5. Electrical codes and standards: Familiarize yourself with the National Electrical Code (NEC) and other relevant industry standards. Learn about common code requirements for wiring, grounding, and overcurrent protection.
6. Renewable energy and sustainability: Understand the basics of renewable energy sources, such as solar, wind, and hydroelectric power. Learn about the environmental benefits of these sources and the challenges associated with integrating them into the electrical grid.

To practice reading comprehension in these areas, find articles, reports, or technical documents related to these topics. As you read, apply active reading techniques, like annotating and summarizing key points. Focus on understanding the author's purpose, the passage's organization, and any technical terms or jargon.

To further reinforce your understanding, discuss these topics with peers or mentors in the electrical industry, or participate in online forums or social media groups. Engaging with others can help you deepen your knowledge and expose you to a variety of perspectives and experiences.

By immersing yourself in electrical industry-related materials and practicing your reading comprehension skills, you'll be better prepared for the IBEW aptitude test's passages.

Here's a passage touching on basic electrical theory: In a far-off corner of an electrical workshop, an apprentice named Lucy marveled at the intricate web of wires and circuits spread out before her. As she delved deeper into the fascinating world of electricity, she began to grasp the significance of basic electrical theory. "Aha!" she exclaimed as the concepts of voltage, current, and resistance started to take shape in her mind.

Voltage, she discovered, is the driving force that pushes electric charge through a conductor. It is like the pressure in a water pipe. Current, on the other hand, represents the flow of electrons, much like the water itself. Resistance, the final piece of this triad, is the material's opposition to the flow of electric current.

In a flash of inspiration, Lucy recalled Ohm's Law: $V = IR$. This fundamental equation connects voltage (V), current (I), and resistance (R) in a linear relationship. Intrigued by this revelation, Lucy pondered how different materials could affect the flow of electricity.

Enter Kirchhoff's Laws, which govern the behavior of electrical circuits. The first law, Kirchhoff's Current Law (KCL), dictates that the total current entering a junction equals the total current leaving it. The second law, Kirchhoff's Voltage Law (KVL), maintains that the sum of the voltages around any closed loop in a circuit must equal zero.

As Lucy delved deeper, she began to appreciate the interconnectedness of these quantities. "Power," she mused, "is the rate at which electrical energy is converted into other forms of energy, like heat or light." $P = IV$, she recalled, connecting power (P) to voltage and current.

Energy, the capacity to do work further solidified these relationships in Lucy's mind. She understood that electrical energy could be calculated by multiplying power by time ($E = Pt$) and marveled at the versatile nature of this invisible force.

With each revelation, Lucy's understanding of basic electrical theory grew, and she felt increasingly confident in her ability to tackle the complex circuits around her. The intricate dance between voltage, current, resistance, and energy now seemed less daunting, and she was eager to continue her journey into the electrifying world of electrical theory.

The sun had just risen, casting a warm glow on the electrical workshop. Eager to continue her journey into the world of electricity, Lucy set out to explore the realm of electrical components and devices. It wasn't long before she encountered a fascinating array of resistors, capacitors, inductors, switches, and transformers.

Resistors, Lucy learned, are crucial components that regulate the flow of electrical current by providing a specific amount of resistance. They come in a variety of shapes and sizes, each with its own unique color-coded bands denoting resistance values. In an electrical circuit, they ensure that sensitive components receive the right amount of current.

Next on her journey, Lucy discovered capacitors – components that store electrical energy in an electric field. She marveled at their ability to release energy when needed and found that they play a pivotal role in filtering, energy storage, and signal coupling. Their symbols, she noted, consist of two parallel lines representing the capacitor plates, sometimes with a curved line to indicate polarity.

As she ventured further, Lucy stumbled upon inductors – coiled components that store energy in a magnetic field. These intriguing devices oppose changes in current flow, and their impact on circuits is described by their inductance, measured in henrys (H). The symbol for an inductor, she observed, is a series of loops or spirals.

Switches, Lucy soon learned, are fundamental components that control the flow of electricity in a circuit. They enable or interrupt the current flow, allowing circuits to be turned on or off as needed. Their symbols, she noted, resemble a bridge with an open or closed gap, indicating whether the switch is open or closed.

Finally, Lucy encountered the remarkable world of transformers. These devices transfer electrical energy between circuits through electromagnetic induction, often changing voltage levels in the process. Consisting of primary and secondary coils wound around a shared core, transformers are depicted by two sets of parallel lines, with the core sometimes represented by vertical lines.

With each new discovery, Lucy's understanding of electrical components and devices grew. She now felt prepared to decipher the language of electrical circuits, the complex interplay of resistors, capacitors, inductors, switches, and transformers, and the powerful symphony they create together.

One day, in a quiet little town, an adventurous electrician named Sam decided to unravel the mysteries of circuit analysis. Equipped with a multimeter, a notepad, and an insatiable curiosity, he set out to explore different techniques, such as series and parallel circuits, Thevenin's theorem, and Norton's theorem.

Sam started by examining series and parallel circuits. He observed that in series circuits, the components were connected end-to-end, with current flowing through each component in turn. The total resistance, he found, was simply the sum of the individual resistances. In parallel circuits, however, components were connected side by side, sharing voltage but not current. Sam quickly realized that calculating the total resistance required a different approach – the reciprocal of the sum of the reciprocals of the individual resistances.

Eager to learn more, Sam delved into Thevenin's theorem, a powerful technique for simplifying complex circuits. He learned that any linear circuit could be reduced to an equivalent circuit with a single voltage source and a series resistor. By removing a load from the original circuit, Sam could calculate the open-circuit voltage and the equivalent resistance, allowing him to easily analyze the behavior of the circuit under varying load conditions.

Next on his journey, Sam encountered Norton's theorem – another method to simplify circuits, but with a twist. Norton's theorem enabled him to transform a linear circuit into an equivalent circuit with a current source in parallel with a resistor. Intrigued by the duality between Thevenin's and Norton's theorems, Sam practiced converting between the two equivalent circuits and found that they provided complementary insights into circuit behavior.

As he honed his skills, Sam tackled increasingly complex problems involving voltages, currents, and resistances in various configurations. His multimeter became an extension of his hand, and his notepad filled with equations, diagrams, and solutions.

Over time, Sam became an expert in circuit analysis. His understanding of series and parallel circuits, Thevenin's theorem, and Norton's theorem allowed him to tackle even the most challenging electrical puzzles. As he looked back on his journey, he couldn't help but feel a sense of pride and accomplishment. His dedication and curiosity had transformed him into a master electrician, ready to illuminate the world with his newfound knowledge.

In the world of electricity, a plethora of **mathematical formulas** are available to describe the intricate relationships among various components and quantities. Here, I've assembled a concise list of essential electrical formulas that you'll likely encounter in your studies, paired with a touch of perplexity and burstiness.

1. Ohm's Law: A cornerstone of electrical theory, Ohm's Law defines the relationship between voltage (V), current (I), and resistance (R) in a circuit. $V = I \times R$
2. Power: The rate at which energy is transferred, used, or transformed, power (P) is calculated using voltage, current, and resistance. $P = V \times I$ $P = I^2 \times R$ $P = V^2 / R$

3. Energy: Representing the capacity to do work, energy (E) is calculated using power and time (t). $E = P \times t$
4. Series circuits: In a series circuit, the total resistance (R_total) equals the sum of individual resistances. $R\_total = R1 + R2 + R3 + ...$
5. Parallel circuits: For parallel circuits, the reciprocal of the total resistance equals the sum of the reciprocals of individual resistances. $1/R\_total = 1/R1 + 1/R2 + 1/R3 + ...$
6. Capacitance: The ability of a capacitor to store an electrical charge, capacitance (C) is calculated using the formula: $C = Q / V$
7. Inductance: A measure of an inductor's ability to oppose changes in current, inductance (L) is given by the formula: $V = L \times (\Delta I / \Delta t)$
8. Impedance: Representing the opposition a circuit offers to alternating current, impedance (Z) is a complex quantity that includes resistance (R) and reactance (X). $Z = \sqrt{R^2 + X^2}$

These formulas serve as a helpful guide to understanding and solving various electrical problems.

Practice Exam Sections:

This section will be the most important part of preparing for you IBEW Test. Studys show time again that the more practice exam questions you do the higher you will score.

To prevent page flipping, we have done a format different from some practice tests. Rather than flipping to an answer key in the back, the answer and explanation will immediately follow the question. To prevent spoiling the answer, take a spare sheet of paper or something similar to hide the answer and prevent accidental peeking. We have found that many students prefer this format instead of flipping through to find the answer key and flipping back and forth.

Please also note that some of the most important topics will be covered several times and some questions will be near repeats of other questions for the ones that are most crucial for you to know for the exam. It may seem repetitive at times, but this is to hammer the most important parts and topics for your exam.

There will be a wide variety of questions, ranging from easy to hard, in no particular order. Good Luck!

A water tank has a capacity of 200 gallons. If it is currently 40% full, how many gallons of water are in the tank?
a) 50 gallons
b) 80 gallons
c) 100 gallons
d) 120 gallons

Answer: b) 80 gallons
Explanation: To find how many gallons are in the tank, multiply the capacity by the percentage full: 200 gallons × 0.4 = 80 gallons.

A transformer has a primary voltage of 480 volts and a secondary voltage of 240 volts. If the primary current is 10 amps, what is the secondary current?
a) 5 amps
b) 10 amps
c) 20 amps
d) 40 amps

Answer: c) 20 amps
Explanation: Use the transformer formula: (Primary Voltage × Primary Current) = (Secondary Voltage × Secondary Current). Plugging in the values: (480 V × 10 A) = (240 V × Secondary Current). Solve for the secondary current: Secondary Current = (480 V × 10 A) / 240 V = 20 A.

A 20-ohm resistor and a 30-ohm resistor are connected in parallel. What is the equivalent resistance?
a) 10 ohms
b) 12 ohms
c) 15 ohms
d) 50 ohms

Answer: b) 12 ohms
Explanation: For parallel resistors, use the formula: 1/Req = 1/R1 + 1/R2. Plug in the values: 1/Req = 1/20 + 1/30. Solve for Req: Req = 1/(1/20 + 1/30) = 12 ohms.

A 12-volt battery is connected to a circuit with a total resistance of 4 ohms. What is the current flowing through the circuit?
a) 2 A
b) 3 A
c) 4 A
d) 8 A

Answer: b) 3 A
Explanation: Use Ohm's Law (V = IR) to find the current: I = V/R. Plugging in the values: I = 12 V / 4 Ω = 3 A.

An electrician needs to install a 30-foot long conduit with a 10% allowance for bends and waste. How many feet of conduit should the electrician order?
a) 30 feet
b) 33 feet
c) 36 feet
d) 39 feet

Answer: b) 33 feet
Explanation: Calculate the allowance by multiplying the length by the percentage: 30 feet × 0.1 = 3 feet. Add the allowance to the original length: 30 feet + 3 feet = 33 feet.

A circuit has a current of 15 amps and a voltage of 120 volts. How much power is being consumed by the circuit?
a) 1,800 watts
b) 2,000 watts
c) 2,500 watts
d) 3,000 watts

Answer: a) 1,800 watts
Explanation: Power is calculated using the formula P = IV. Plugging in the values: P = 15 A × 120 V = 1,800 W.

An electrician needs to wire a room with 8 outlets, each requiring 2 feet of wire. If the electrician has a 100-foot spool of wire, how many feet of wire will be left after wiring the outlets?
a) 68 feet
b) 76 feet
c) 84 feet
d) 92 feet

Answer: c) 84 feet
Explanation: Calculate the total wire needed for the outlets: 8 outlets × 2 feet = 16 feet. Subtract the total wire needed from the spool length: 100 feet - 16 feet = 84 feet.

A motor draws 50 amps of current and has a power factor of 0.9. What is the apparent power in volt-amperes (VA)?
a) 4,500 VA
b) 5,000 VA
c) 5,500 VA
d) 6,000 VA

Answer: a) 4,500 VA
Explanation: Apparent power (S) is calculated using the formula S = I / PF. Plugging in the values: S = 50 A / 0.9 = 4,500 VA.

A 240-volt circuit has a resistance of 60 ohms. What is the current flowing through the circuit?
a) 1 A
b) 2 A
c) 4 A
d) 6 A

Answer: c) 4 A
Explanation: Use Ohm's Law (V = IR) to find the current: I = V/R. Plugging in the values: I = 240 V / 60 Ω = 4 A.

A 200-watt light bulb is connected to a 120-volt power source. How many amps of current does the light bulb draw?
a) 0.6 A
b) 1.2 A
c) 1.6 A
d) 1.8 A

Answer: d) 1.8 A
Explanation: Use the formula P = IV to find the current: I = P/V. Plugging in the values: I = 200 W / 120 V = 1.8 A.

A transformer has a primary winding with 1,000 turns and a secondary winding with 500 turns. If the primary voltage is 400 volts, what is the secondary voltage?
a) 100 volts
b) 200 volts
c) 300 volts
d) 400 volts

Answer: b) 200 volts
Explanation: Use the turns ratio formula: V_primary / V_secondary = N_primary / N_secondary. Solve for V_secondary: V_secondary = (V_primary × N_secondary) / N_primary = (400 V × 500 turns) / 1000 turns = 200 V.

An electrician is installing wires in a building. If he installs 12 feet of wire in 10 minutes, how many feet of wire can he install in 1 hour?
a) 36 feet
b) 72 feet
c) 144 feet
d) 216 feet

Answer: b) 72 feet
Explanation: Find the proportion of wire installed per minute: 12 feet / 10 minutes = 1.2 feet/minute. Multiply this rate by the number of minutes in an hour (60 minutes) to find the total wire installed in an hour: 1.2 feet/minute × 60 minutes = 72 feet.

A circuit with a power factor of 0.8 has an apparent power of 1,000 volt-amperes (VA). What is the circuit's real power in watts?
a) 200 W
b) 500 W
c) 800 W
d) 1,000 W

Answer: c) 800 W
Explanation: Use the power factor formula: Real power = Apparent power × Power factor. Plugging in the values: 1,000 VA × 0.8 = 800 W.

In a 3-phase system, the ratio of line voltage to phase voltage for a delta configuration is 1:1. If the line voltage is 480 volts, what is the phase voltage?
a) 120 volts
b) 240 volts
c) 480 volts
d) 960 volts

Answer: c) 480 volts
Explanation: In a delta configuration, line voltage and phase voltage are equal. Therefore, if the line voltage is 480 volts, the phase voltage is also 480 volts.

A 60-watt light bulb operates at 120 volts. If an electrician needs to install a 240-volt light bulb with the same power rating, what should be the resistance of the new bulb?
a) 120 ohms
b) 240 ohms
c) 480 ohms
d) 960 ohms

Answer: d) 960 ohms
Explanation: First, find the current for the 60-watt, 120-volt light bulb using P = IV. I = P/V = 60 W / 120 V = 0.5 A. Then, find the resistance using Ohm's Law (V = IR): R = V/I = 120 V / 0.5 A = 240 ohms. Since power is constant, the new bulb at 240 volts will draw half the current (0.25 A). Find the resistance for the new bulb using Ohm's Law: R = V/I = 240 V / 0.25 A = 960 ohms.

A 50-kilowatt (kW) generator operates at 80% efficiency. How much power is actually generated and available for use?
a) 20 kW
b) 40 kW
c) 50 kW
d) 62.5 kW

Answer: b) 40 kW
Explanation: Multiply the rated power by the efficiency: 50 kW × 80% = 40 kW.

A transformer increases the voltage by 200%. If the input voltage is 120 volts, what is the output voltage?
a) 240 volts
b) 360 volts
c) 480 volts
d) 600 volts

Answer: c) 480 volts
Explanation: Multiply the input voltage by 1 plus the percentage increase: 120 V × (1 + 200%) = 120 V × 3 = 480 V.

An electrician notices that a 20-ampere (A) circuit breaker is tripping when the load reaches 18 A. What percentage of the circuit breaker's rated current is causing the breaker to trip?
a) 75%
b) 80%
c) 90%
d) 95%

Answer: c) 90%
Explanation: Divide the tripping current by the rated current and multiply by 100%: (18 A / 20 A) × 100% = 90%.

A motor's efficiency is 75%. If the motor consumes 6,000 watts of power, how much power is converted to useful work?
a) 1,500 watts
b) 3,000 watts
c) 4,500 watts
d) 6,000 watts

Answer: c) 4,500 watts
Explanation: Multiply the consumed power by the efficiency: 6,000 W × 75% = 4,500 W.

An electrical installation requires a 5% voltage drop to maintain proper operation. If the initial voltage is 240 volts, what is the allowable voltage drop in volts?
a) 5 volts
b) 10 volts
c) 12 volts
d) 15 volts

Answer: c) 12 volts
Explanation: Multiply the initial voltage by the percentage of the voltage drop: 240 V × 5% = 12 V.

A capacitor has a capacitance of 16 microfarads (μF). If the capacitance is doubled, what is the square root of the new capacitance?
a) 4 μF
b) 8 μF
c) 16 μF
d) 32 μF

Answer: b) 8 μF
Explanation: Double the capacitance: 16 μF × 2 = 32 μF. The square root of 32 μF is 8 μF.

The resistance in an electrical circuit follows the formula R = V^2 / P, where R is resistance, V is voltage, and P is power. If the voltage is 120 volts and the power is 1,440 watts, what is the resistance?
a) 10 ohms
b) 20 ohms
c) 30 ohms
d) 40 ohms

Answer: a) 10 ohms
Explanation: R = (120 V)^2 / 1,440 W = 14,400 / 1,440 = 10 ohms.

A 500-watt (W) device uses a power formula P = IV, where P is power, I is current, and V is voltage. If the voltage is 100 volts, what is the cube root of the current?
a) 1 A
b) 2 A
c) 3 A
d) 5 A

Answer: a) 1 A
Explanation: Rearrange the formula to solve for I: I = P / V. I = 500 W / 100 V = 5 A. The cube root of 5 A is approximately 1 A.

An electrical circuit has a power factor of 0.8. If the apparent power is 250 volt-amperes (VA), what is the square of the real power?
a) 16,000 W²
b) 25,000 W²
c) 36,000 W²
d) 49,000 W²

Answer: c) 36,000 W²
Explanation: Multiply the apparent power by the power factor to find the real power: 250 VA × 0.8 = 200 W. The square of the real power is 200² = 36,000 W².

The total resistance of a parallel circuit is calculated using the formula 1/Rt = 1/R1 + 1/R2, where Rt is the total resistance, R1 and R2 are the individual resistances. If R1 is 4 ohms and R2 is 6 ohms, what is the square root of the total resistance?
a) 1 ohm
b) 2 ohms
c) 3 ohms
d) 4 ohms

Answer: b) 2 ohms
Explanation: 1/Rt = 1/4 + 1/6 = 3/12 + 2/12 = 5/12. Therefore, Rt = 12/5 or 2.4 ohms. The square root of 2.4 ohms is approximately 2 ohms.

An electrician needs to calculate the current in a circuit with a resistance of R ohms and a voltage of V volts. The formula for current is I = V/R. If V = 3R, what is the current in terms of R?
a) I = R/3
b) I = 3/R
c) I = 3
d) I = R

Answer: b) I = 3/R
Explanation: Replace V with 3R in the current formula: I = (3R)/R. Simplify: I = 3/R.

A circuit has a power of P watts, a voltage of V volts, and a current of I amperes. The power formula is P = IV. If P = 2V and I = 3A, what is the voltage in the circuit?
a) V = 2 volts
b) V = 3 volts
c) V = 4 volts
d) V = 6 volts

Answer: d) V = 6 volts
Explanation: Use the power formula P = IV. Replace P with 2V: 2V = 3A * V. Divide both sides by 3A: V = 2/3 * 2V. Solve: V = 6 volts.

An electrician is installing three resistors in a series circuit. The resistors have values of R1, R2, and R3 ohms. The total resistance in a series circuit is the sum of individual resistances. If R1 = 3x, R2 = 2x, and R3 = x, what is the total resistance in terms of x?
a) 6x
b) 5x
c) 4x
d) 3x

Answer: a) 6x
Explanation: The total resistance is Rt = R1 + R2 + R3. Substitute the values: Rt = 3x + 2x + x. Simplify: Rt = 6x.

A transformer has a primary coil with N1 turns and a secondary coil with N2 turns. The voltage across the primary coil is V1, and the voltage across the secondary coil is V2. The transformer follows the formula V1/V2 = N1/N2. If V1 = 120 volts, N1 = 240 turns, and N2 = 480 turns, what is the voltage across the secondary coil?
a) 30 volts
b) 60 volts
c) 90 volts
d) 120 volts

Answer: b) 60 volts
Explanation: Use the transformer formula: V1/V2 = N1/N2. Substitute the values: 120/V2 = 240/480. Simplify and solve: V2 = 60 volts.

An electrician needs to calculate the power (P) in a circuit with a current (I) and a resistance (R). The power formula is P = I^2 * R. If I = 4A and R = 5 ohms, what is the power in the circuit?
a) 20 watts
b) 40 watts
c) 80 watts
d) 100 watts

Answer: c) 80 watts
Explanation: Substitute the values into the power formula: P = (4A)^2 * 5 ohms. Calculate: P = 16A² * 5 ohms = 80 watts.

An electrician needs to determine the length of wire required to run along the diagonal of a rectangular room. The room measures 20 feet by 15 feet. What is the length of the diagonal wire?
a) 22 feet
b) 25 feet
c) 28 feet
d) 35 feet

Answer: b) 25 feet
Explanation: Use the Pythagorean theorem ($a^2 + b^2 = c^2$) with a = 20 feet and b = 15 feet: ($20^2 + 15^2$) = c^2. Calculate: (400 + 225) = c^2. Solve: c = 25 feet.

A right-angled triangle has a hypotenuse measuring 13 inches and one leg measuring 5 inches. What is the length of the other leg?
a) 8 inches
b) 10 inches
c) 12 inches
d) 14 inches

Answer: c) 12 inches
Explanation: Use the Pythagorean theorem ($a^2 + b^2 = c^2$) with the hypotenuse c = 13 inches and one leg a = 5 inches: ($5^2 + b^2$) = 13^2. Calculate: (25 + b^2) = 169. Solve: b = 12 inches.

An electrician needs to find the area of a circular conduit with a diameter of 10 inches. What is the area of the conduit's cross-section?
a) 25π square inches
b) 50π square inches
c) 75π square inches
d) 100π square inches

Answer: d) 100π square inches
Explanation: The radius of the circle is half the diameter, so r = 5 inches. Use the area formula A = πr^2: A = π(5^2) = 100π square inches.

An electrician must find the volume of a cylindrical wire spool that is 6 inches high and has a diameter of 4 inches. What is the volume of the spool in cubic inches?
a) 48π cubic inches
b) 72π cubic inches
c) 96π cubic inches
d) 120π cubic inches

Answer: b) 72π cubic inches
Explanation: The radius of the cylinder is half the diameter, so r = 2 inches. Use the volume formula V = πr²h: V = π(2²)(6) = 72π cubic inches.

A triangular trench with a right angle has a base of 30 meters and a height of 40 meters. What is the area of the triangular trench?
a) 500 square meters
b) 600 square meters
c) 700 square meters
d) 800 square meters

Answer: b) 600 square meters
Explanation: Use the area formula for a triangle A = 1/2 * base * height: A = 1/2 * 30m * 40m = 600 square meters.

An electrician installs a light fixture at point A(4, 6) and a switch at point B(-2, 6). What is the horizontal distance between the light fixture and the switch?
a) 2 units
b) 4 units
c) 6 units
d) 8 units

Answer: d) 8 units
Explanation: The horizontal distance between two points is calculated by subtracting the x-coordinates: 4 - (-2) = 4 + 2 = 8 units.

A circuit board has two components placed at coordinates P(3, -5) and Q(-3, 5). What is the distance between the two components?
a) 5 units
b) 10 units
c) 5√2 units
d) 10√2 units

Answer: d) 10√2 units
Explanation: Use the distance formula: $\sqrt{(x_2 - x_1)^2 + (y_2 - y_1)^2}$. Calculate: $\sqrt{(-3 - 3)^2 + (5 - (-5))^2} = \sqrt{(-6)^2 + 10^2} = \sqrt{36 + 100} = \sqrt{136} = 10\sqrt{2}$ units.

A power outlet is installed at point A(5, 3), and a straight wire runs from point A to point B(7, 9). What is the slope of the wire?
a) 1/2
b) 2
c) 3
d) 4

Answer: c) 3
Explanation: Calculate the slope (m) using the formula $m = (y_2 - y_1) / (x_2 - x_1)$: m = (9 - 3) / (7 - 5) = 6 / 2 = 3.

An electrician needs to find the midpoint of a cable that connects two points, A(-4, 2) and B(6, 10). What are the coordinates of the midpoint?
a) (1, 6)
b) (1, 4)
c) (2, 6)
d) (2, 4)

Answer: a) (1, 6)
Explanation: Use the midpoint formula: $((x_1 + x_2) / 2, (y_1 + y_2) / 2)$. Calculate: ((-4 + 6) / 2, (2 + 10) / 2) = (1, 6).

A transformer is located at point T(3, -2). The electrician needs to find the point S(x, y) such that the slope of the line between the points is 2 and the x-coordinate of point S is 5. What are the coordinates of point S?
a) (5, -6)
b) (5, 0)
c) (5, 2)
d) (5, 4)

Answer: d) (5, 4)
Explanation: Use the slope formula: m = (y2 - y1) / (x2 - x1). With m = 2, x1 = 3, y1 = -2, and x2 = 5, we get 2 = (y - (-2)) / (5 - 3). Solve for y: 4 = y + 2, y = 4. So, the coordinates of point S are (5, 4).

An electrician installs a 300-watt light bulb that is used for 4 hours every day. How much energy does the light bulb consume in a 30-day month?
a) 3,600 watt-hours
b) 12,000 watt-hours
c) 36,000 watt-hours
d) 120,000 watt-hours

Answer: c) 36,000 watt-hours
Explanation: Multiply the wattage by the daily usage and the number of days: 300 watts * 4 hours/day * 30 days = 36,000 watt-hours.

A 120-volt circuit has a 20-amp circuit breaker. What is the maximum power that can be safely drawn from the circuit?
a) 100 watts
b) 600 watts
c) 1,200 watts
d) 2,400 watts

Answer: d) 2,400 watts
Explanation: Use the formula P = V * I: P = 120 volts * 20 amps = 2,400 watts.

A transformer has a primary voltage of 480 volts and a secondary voltage of 120 volts. If the primary coil has 800 turns, how many turns are there in the secondary coil?
a) 200 turns
b) 400 turns
c) 600 turns
d) 800 turns

Answer: a) 200 turns
Explanation: Use the formula Vp/Vs = Np/Ns: 480 volts / 120 volts = 800 turns / Ns. Solve for Ns: Ns = 800 turns * (120/480) = 200 turns.

A 50-foot wire has a resistance of 0.1 ohms. If an electrician needs to use a 150-foot wire with the same resistance per foot, what will be the total resistance of the longer wire?
a) 0.1 ohms
b) 0.3 ohms
c) 0.5 ohms
d) 1 ohm

Answer: b) 0.3 ohms
Explanation: Divide the total resistance by the length of the shorter wire to get the resistance per foot: 0.1 ohms / 50 feet = 0.002 ohms/foot. Multiply by the length of the longer wire: 0.002 ohms/foot * 150 feet = 0.3 ohms.

An electrician charges $45 per hour for labor and $75 for a service call fee. If the electrician works for 3 hours on a job, what will the total cost be?
a) $135
b) $210
c) $240
d) $300

Answer: b) $210
Explanation: Multiply the hourly rate by the hours worked and add the service call fee: ($45/hour * 3 hours) + $75 = $135 + $75 = $210.

An electrician purchases 25 feet of wire at $0.50 per foot and 12 feet of conduit at $2.00 per foot. How much does the electrician spend on these materials?
a) $16.50
b) $24.00
c) $29.00
d) $49.00

Answer: d) $49.00
Explanation: Calculate the total cost of wire and conduit: (25 feet * $0.50/foot) + (12 feet * $2.00/foot) = $12.50 + $24.00 = $49.00.

An electrical panel has three circuits. Circuit 1 has a 15-amp breaker, circuit 2 has a 20-amp breaker, and circuit 3 has a 25-amp breaker. What is the total amperage of the panel?
a) 45 amps
b) 50 amps
c) 55 amps
d) 60 amps

Answer: c) 55 amps
Explanation: Add the amperages of the three circuits: 15 amps + 20 amps + 25 amps = 55 amps.

A warehouse has 6 rows of lights with 8 lights in each row. If each light uses 75 watts, how many total watts are used by all the lights in the warehouse?
a) 300 watts
b) 3600 watts
c) 4800 watts
d) 36000 watts

Answer: b) 3600 watts
Explanation: Calculate the total number of lights: 6 rows * 8 lights/row = 48 lights. Multiply by the wattage per light: 48 lights * 75 watts/light = 3600 watts.

An electrician installs 7 switches at a cost of $3.50 each and 5 outlets at a cost of $2.50 each. How much does the electrician spend on switches and outlets?
a) $24.50
b) $29.50
c) $33.00
d) $36.50

Answer: a) $24.50
Explanation: Calculate the total cost of switches and outlets: (7 switches * $3.50/switch) + (5 outlets * $2.50/outlet) = $24.50 + $12.50 = $24.50.

A building has 4 floors with 20 electrical outlets per floor. Each electrical outlet has a maximum current of 15 amps. What is the total current capacity of all outlets in the building?
a) 60 amps
b) 240 amps
c) 600 amps
d) 1200 amps

Answer: d) 1200 amps
Explanation: Calculate the total number of outlets: 4 floors * 20 outlets/floor = 80 outlets. Multiply by the maximum current per outlet: 80 outlets * 15 amps/outlet = 1200 amps.

An electrician initially has 200 feet of wire. After using 135 feet for a project, how many feet of wire does the electrician have left?
a) 55 feet
b) 65 feet
c) 75 feet
d) 85 feet

Answer: b) 65 feet
Explanation: Subtract the amount of wire used from the initial amount: 200 feet - 135 feet = 65 feet.

A circuit has a total resistance of 150 ohms. If one resistor has a resistance of 90 ohms, what is the resistance of the other resistor in the circuit?
a) 50 ohms
b) 60 ohms
c) 65 ohms
d) 75 ohms

Answer: b) 60 ohms
Explanation: Subtract the resistance of the known resistor from the total resistance: 150 ohms - 90 ohms = 60 ohms.

A 480-volt electrical system has a voltage drop of 80 volts across a conductor. What is the remaining voltage after the drop?
a) 360 volts
b) 400 volts
c) 420 volts
d) 440 volts

Answer: b) 400 volts
Explanation: Subtract the voltage drop from the initial voltage: 480 volts - 80 volts = 400 volts.

A 100-amp service has a load of 85 amps. How many additional amps are available for the service?
a) 5 amps
b) 10 amps
c) 15 amps
d) 20 amps

Answer: c) 15 amps
Explanation: Subtract the load from the service capacity: 100 amps - 85 amps = 15 amps.

An electrical system has a 60 Hz frequency. If the frequency drops by 8 Hz, what is the new frequency of the system?
a) 48 Hz
b) 52 Hz
c) 56 Hz
d) 62 Hz

Answer: b) 52 Hz
Explanation: Subtract the frequency drop from the initial frequency: 60 Hz - 8 Hz = 52 Hz.

An electrician needs to install six 15-amp breakers in a panel. What is the total amperage capacity of these breakers?
a) 60 amps
b) 75 amps
c) 90 amps
d) 105 amps

Answer: c) 90 amps
Explanation: Multiply the number of breakers by the amperage of each: 6 breakers * 15 amps = 90 amps.

A 120-volt circuit has a current of 10 amps. What is the power consumption of the circuit in watts?
a) 1,200 watts
b) 1,100 watts
c) 1,000 watts
d) 900 watts

Answer: a) 1,200 watts
Explanation: Multiply the voltage by the current: 120 volts * 10 amps = 1,200 watts.

A transformer has a primary voltage of 240 volts and a secondary voltage of 120 volts. If the primary winding has 200 turns, how many turns are in the secondary winding?
a) 50 turns
b) 100 turns
c) 150 turns
d) 200 turns

Answer: b) 100 turns
Explanation: Calculate the ratio of the primary voltage to the secondary voltage: 240 volts / 120 volts = 2. Multiply the primary turns by the inverse of the ratio: 200 turns / 2 = 100 turns.

A device has a power rating of 60 watts and is connected to a 120-volt supply. What is the current draw of the device in amps?
a) 0.5 amps
b) 1 amp
c) 2 amps
d) 4 amps

Answer: a) 0.5 amps
Explanation: Divide the power rating by the supply voltage: 60 watts / 120 volts = 0.5 amps.

A motor has a power factor of 0.8 and a real power of 400 watts. What is the apparent power in volt-amperes (VA)?
a) 320 VA
b) 500 VA
c) 600 VA
d) 800 VA

Answer: b) 500 VA
Explanation: Divide the real power by the power factor: 400 watts / 0.8 = 500 VA.

An electrical contractor needs to divide a 3/4-inch conduit into five equal sections. What is the length of each section?
a) 1/8 inch
b) 1/4 inch
c) 3/8 inch
d) 1/2 inch

Answer: c) 3/8 inch
Explanation: Divide the length of the conduit by the number of sections: (3/4) / 5 = 3/20. Convert the fraction to a more familiar form: 3/20 = 3/8 inch.

A 1,200-watt heater operates for 8 hours a day. If the cost of electricity is $0.15 per kilowatt-hour, how much does it cost to operate the heater for one day?
a) $1.44
b) $2.88
c) $5.76
d) $11.52

Answer: a) $1.44
Explanation: Convert watts to kilowatts: 1,200 watts = 1.2 kW. Multiply the kilowatts by the hours of operation: 1.2 kW * 8 hours = 9.6 kWh. Multiply the energy consumption by the cost per kilowatt-hour: 9.6 kWh * $0.15 = $1.44.

An electrician completes 12 out of 15 tasks on their list. What percentage of tasks has the electrician completed?
a) 60%
b) 70%
c) 75%
d) 80%

Answer: d) 80%
Explanation: Divide the number of completed tasks by the total number of tasks: 12 / 15 = 0.8. Convert the decimal to a percentage: 0.8 * 100 = 80%.

A motor operates at 80% efficiency. If the input power is 1,500 watts, what is the output power?
a) 1,000 watts
b) 1,200 watts
c) 1,350 watts
d) 1,500 watts

Answer: b) 1,200 watts
Explanation: Multiply the input power by the efficiency: 1,500 watts * 0.8 = 1,200 watts.

A transformer has an efficiency of 95%. If the output power is 3,800 watts, what is the input power?
a) 3,610 watts
b) 3,800 watts
c) 3,995 watts
d) 4,000 watts

Answer: d) 4,000 watts
Explanation: Divide the output power by the efficiency: 3,800 watts / 0.95 = 4,000 watts.

An electrical panel has a ratio of 3 circuit breakers to 2 fuses. If there are 18 circuit breakers in the panel, how many fuses are present?
a) 8
b) 10
c) 12
d) 14

Answer: c) 12
Explanation: Divide the number of circuit breakers by the ratio of circuit breakers: 18 / 3 = 6. Multiply this result by the ratio of fuses: 6 * 2 = 12 fuses.

An electrician has to install 15 outlets for every 5 light switches in a building. If the electrician installs 45 light switches, how many outlets need to be installed?
a) 60
b) 90
c) 135
d) 180

Answer: d) 180
Explanation: Divide the number of light switches by the ratio of light switches: 45 / 5 = 9. Multiply this result by the ratio of outlets: 9 * 15 = 180 outlets.

A transformer has a primary-to-secondary turns ratio of 4:1. If the primary coil has 200 turns, how many turns are in the secondary coil?
a) 50
b) 100
c) 150
d) 200

Answer: a) 50
Explanation: Divide the primary turns by the primary-to-secondary turns ratio: 200 turns / 4 = 50 turns.

An electrical device has a voltage-to-current ratio of 7:3. If the current is 12 amps, what is the voltage?
a) 24 V
b) 28 V
c) 36 V
d) 56 V

Answer: d) 56 V
Explanation: Divide the current by the current part of the ratio: 12 amps / 3 = 4. Multiply this result by the voltage part of the ratio: 4 * 7 = 56 V.

The ratio of the number of hours an electrician works on residential projects to commercial projects is 5:3. If the electrician works 32 hours on residential projects in a week, how many hours do they work on commercial projects?
a) 12
b) 16
c) 20
d) 24

Answer: d) 24
Explanation: Divide the number of residential hours by the residential part of the ratio: 32 hours / 5 = 6.4. Multiply this result by the commercial part of the ratio: 6.4 * 3 = 19.2 (rounded to the nearest whole number, 24).

An electrician is calculating the total resistance (R) of a parallel circuit with two resistors, R1 and R2. The formula for total resistance in a parallel circuit is 1/R = 1/R1 + 1/R2. If R1 = 6 ohms and R2 = 3 ohms, what is the total resistance R?
a) 1 ohm
b) 2 ohms
c) 3 ohms
d) 4 ohms

Answer: b) 2 ohms
Explanation: Using the formula, 1/R = 1/6 + 1/3 = 1/6 + 2/6 = 3/6. To find R, take the reciprocal: R = 6/3 = 2 ohms.

A motor's power (P) is given by the formula P = IV, where I is the current and V is the voltage. If the power is 900 watts and the current is 10 amps, what is the voltage?
a) 90 V
b) 100 V
c) 110 V
d) 120 V

Answer: a) 90 V
Explanation: Rearrange the formula to solve for V: V = P/I. Substitute the given values: V = 900 W / 10 A = 90 V.

An electrician has a linear equation that models the relationship between the number of hours worked (x) and the total amount earned (y). If the electrician earns $50 for 3 hours of work and $100 for 6 hours of work, what is the equation?
a) y = 16.67x
b) y = 25x
c) y = 33.33x
d) y = 50x

Answer: b) y = 25x
Explanation: Calculate the slope (m) using the two points: m = (100 - 50) / (6 - 3) = 50 / 3 = 16.67. This is the amount earned per hour. Using either point, find the equation: y = 16.67 * x.

A capacitor's time constant (τ) is given by the formula τ = RC, where R is the resistance and C is the capacitance. If the time constant is 4 seconds and the resistance is 8 ohms, what is the capacitance?
a) 0.25 F
b) 0.5 F
c) 1 F
d) 2 F

Answer: c) 1 F
Explanation: Rearrange the formula to solve for C: C = τ/R. Substitute the given values: C = 4 s / 8 Ω = 0.5 F.

If a series circuit has a total resistance (R) of 20 ohms and the sum of the individual resistances is represented by the equation R = 3x + 5, what is the value of x?
a) 3
b) 4
c) 5
d) 6

Answer: d) 6
Explanation: Substitute the given value for R into the equation: 20 = 3x + 5. Solve for x: 15 = 3x, x = 5.

Solar power, hydrogen fuel cells, and compressed natural gas are examples of alternative energy sources that could help reduce the United States' reliance on oil, as 94% of transportation relies on it, and 84.3% is imported. The potential for alternative energy to mitigate the environmental impact of transportation is also recognized, but high costs and insufficient infrastructure, such as charging and refueling stations, hinder consumer adoption of green vehicles. The Department of Transportation contends that assistance is necessary for the widespread adoption of alternative energy vehicles and has suggested a set of mandatory goals for states to establish a baseline level of clean fuel infrastructure. This would provide consumers with the confidence to purchase vehicles powered by non-fossil fuels.

What is the most probable consequence if the Department of Transportation's proposal is put into effect?

a) An increase in the availability of charging and refueling stations for alternative energy vehicles.
b) A sudden and significant decrease in oil prices.
c) The immediate replacement of all fossil fuel-powered vehicles.
d) The total elimination of the environmental impact caused by transportation.

Answer:
a) An increase in the availability of charging and refueling stations for alternative energy vehicles.

Reading Passage:

Deforestation is a significant environmental issue affecting many regions across the globe. By cutting down trees and destroying forests, humans are disrupting natural habitats and contributing to climate change. Forests are critical for maintaining the balance of carbon dioxide in the atmosphere, as they absorb and store this greenhouse gas. In addition, forests provide shelter and resources for countless species of plants and animals. Conservation efforts are crucial in order to preserve these vital ecosystems and protect biodiversity. Sustainable forestry practices, reforestation programs, and public awareness campaigns are just a few examples of strategies that can be employed to combat deforestation.

Question 1:
What is the primary issue discussed in the passage?

A. Climate change
B. Deforestation
C. Sustainable forestry practices
D. Biodiversity

Answer: B. Deforestation

Question 2:
What role do forests play in maintaining the balance of carbon dioxide in the atmosphere?

A. They release carbon dioxide.
B. They absorb and store carbon dioxide.
C. They have no effect on carbon dioxide levels.
D. They convert carbon dioxide into other greenhouse gases.

Answer: B. They absorb and store carbon dioxide.

Question 3:
Which of the following is NOT mentioned as a strategy to combat deforestation in the passage?

A. Sustainable forestry practices
B. Reforestation programs
C. Public awareness campaigns
D. Increasing greenhouse gas emissions

Answer: D. Increasing greenhouse gas emissions

Reading Passage:

Solar energy is a renewable energy source harnessed from the sun's rays. It has the potential to provide a significant portion of the world's energy needs, reducing the reliance on fossil fuels and decreasing greenhouse gas emissions. There are two primary methods for harnessing solar power: photovoltaic (PV) cells and concentrating solar power (CSP) systems. Photovoltaic cells directly convert sunlight into electricity, whereas CSP systems use mirrors to concentrate sunlight onto a focal point, generating heat that is used to produce electricity. Both technologies have their advantages and disadvantages, but widespread adoption of solar energy is essential for a sustainable future.

Question 1:
What is the main topic of the passage?

A. Greenhouse gas emissions
B. Fossil fuels
C. Renewable energy sources
D. Solar energy

Answer: D. Solar energy

Question 2:
What are the two primary methods for harnessing solar power mentioned in the passage?

A. Solar panels and wind turbines
B. Photovoltaic cells and concentrating solar power systems
C. Biomass and hydroelectric power
D. Geothermal energy and tidal power

Answer: B. Photovoltaic cells and concentrating solar power systems

Question 3:
Why is the widespread adoption of solar energy essential?

A. It is the only renewable energy source.
B. It can reduce reliance on fossil fuels and decrease greenhouse gas emissions.
C. It is the cheapest form of energy.
D. Solar energy can only be harnessed in specific locations.

Answer: B. It can reduce reliance on fossil fuels and decrease greenhouse gas emissions.

Reading Passage:

Artificial intelligence (AI) is a rapidly growing field that focuses on the creation of machines capable of intelligent behavior. AI has various applications in numerous industries, including healthcare, finance, and automotive. Machine learning, a subset of AI, involves training computer algorithms to learn from data and make predictions or decisions. Deep learning, a further subset of machine learning, uses neural networks to process vast amounts of data, allowing machines to recognize complex patterns and make more sophisticated decisions. As AI technology advances, it is expected to have a transformative impact on society, changing the way we work, live, and interact with one another.

Question 1:
What is the main focus of artificial intelligence?

A. Manufacturing machinery
B. Creating machines capable of intelligent behavior
C. Developing new computer hardware
D. Enhancing human intelligence

Answer: B. Creating machines capable of intelligent behavior

Question 2:
Which of the following is a subset of AI that trains computer algorithms to learn from data?

A. Deep learning
B. Neural networks
C. Machine learning
D. Data processing

Answer: C. Machine learning

Question 3:
How is AI expected to impact society in the future?

A. It will have no significant impact on society.
B. It will only affect specific industries, such as healthcare and finance.
C. It will change the way we work, live, and interact with one another.
D. It will lead to the decline of technology innovation.

Answer: C. It will change the way we work, live, and interact with one another.

Reading Passage:

Urban agriculture is an innovative movement that seeks to cultivate, process, and distribute food within urban environments. By growing food in cities, proponents of urban agriculture aim to reduce the distance food must travel, thereby reducing its environmental footprint. This movement encourages the repurposing of underutilized urban spaces, such as vacant lots and rooftops, into productive green areas. Urban agriculture promotes local food production, increases access to fresh produce, and fosters community engagement. As urban populations continue to grow, urban agriculture is becoming an increasingly important strategy for addressing food security and sustainability.

Question 1:
What is the primary goal of urban agriculture?

A. To create more green spaces for aesthetic purposes
B. To reduce the environmental footprint of food by growing it in cities
C. To develop new agricultural technologies
D. To increase the number of jobs in agriculture

Answer: B. To reduce the environmental footprint of food by growing it in cities

Question 2:
Which of the following is an example of underutilized urban space that could be repurposed for urban agriculture?

A. Forests
B. Rivers
C. Rooftops
D. Highways

Answer: C. Rooftops

Question 3:
How does urban agriculture contribute to addressing food security and sustainability?

A. By importing food from other countries
B. By promoting local food production and increasing access to fresh produce
C. By increasing reliance on large-scale industrial agriculture
D. By reducing the overall demand for food

Answer: B. By promoting local food production and increasing access to fresh produce

Reading Passage:

Ocean currents play a crucial role in regulating Earth's climate by redistributing heat from the equator to the poles. These vast systems of flowing water are driven by various factors, including the planet's rotation, differences in water temperature, and salinity. The Gulf Stream, a powerful ocean current, transports warm water from the Gulf of Mexico across the Atlantic Ocean to Western Europe. It significantly impacts the climate of the eastern United States and Western Europe by moderating temperature fluctuations. Researchers are closely monitoring ocean currents to better understand the potential effects of climate change on these important systems.

Question 1:
What is the primary function of ocean currents in Earth's climate system?

A. Producing electricity for coastal cities
B. Regulating sea levels around the world
C. Redistributing heat from the equator to the poles
D. Preventing the formation of hurricanes

Answer: C. Redistributing heat from the equator to the poles

Question 2:
Which of the following factors is NOT responsible for driving ocean currents?

A. The Earth's rotation
B. Differences in water temperature
C. The position of the moon
D. Differences in water salinity

Answer: C. The position of the moon

Question 3:
How does the Gulf Stream affect the climate of the eastern United States and Western Europe?

A. By causing more frequent storms
B. By increasing the likelihood of droughts
C. By moderating temperature fluctuations
D. By enhancing coastal erosion

Answer: C. By moderating temperature fluctuations

Reading Passage:

The honeybee plays a crucial role in pollinating plants and ensuring the continuation of diverse ecosystems. As they forage for nectar, these industrious insects transfer pollen from one flower to another, allowing plants to produce seeds and reproduce. Unfortunately, honeybee populations have been declining due to factors such as habitat loss, pesticide exposure, and disease. The decline in honeybee populations poses a threat to global food production, as they are responsible for pollinating a significant portion of the world's crops. Many researchers, farmers, and environmentalists are working together to understand and address the challenges facing honeybees, aiming to protect these valuable pollinators.

Question 1:
What is the primary role of honeybees in the ecosystem?

A. Producing honey for human consumption
B. Feeding on small insects
C. Pollinating plants and supporting their reproduction
D. Controlling the population of other insects

Answer: C. Pollinating plants and supporting their reproduction

Question 2:
Which of the following factors is NOT contributing to the decline in honeybee populations?

A. Habitat loss
B. Pesticide exposure
C. Predation by large mammals
D. Disease

Answer: C. Predation by large mammals

Question 3:
Why is the decline in honeybee populations a concern for global food production?

A. Honeybees are a primary food source for many animals
B. Honey is a staple food in many countries
C. Honeybees pollinate a significant portion of the world's crops
D. The decline in honeybee populations affects the production of honey-based medicines

Answer: C. Honeybees pollinate a significant portion of the world's crops

Reading Passage:

Coral reefs are often referred to as the rainforests of the sea due to their extraordinary biodiversity. These underwater ecosystems are composed of coral polyps, which are tiny animals that secrete a hard, limestone skeleton. Over time, these structures build up, creating a habitat for countless marine species. Coral reefs are not only vital for marine life, but they also offer numerous benefits to humans, including coastal protection, tourism, and medicine. Unfortunately, factors such as climate change, pollution, and overfishing have put coral reefs under threat. Conservation efforts have been implemented worldwide to protect these vital ecosystems and ensure their survival for future generations.

Question 1:
Why are coral reefs often referred to as the rainforests of the sea?

A. They are found in the same regions as rainforests
B. They have a similar structure to rainforests
C. They provide habitat for a wide variety of marine species
D. Both are exclusively found in tropical environments

Answer: C. They provide habitat for a wide variety of marine species

Question 2:
What is the primary building block of coral reefs?

A. Algae
B. Seaweed
C. Coral polyps
D. Sponges

Answer: C. Coral polyps

Question 3:
Which of the following is NOT a benefit that coral reefs provide to humans?

A. Coastal protection from storms and erosion
B. Tourism and recreation opportunities
C. Sources of new medicines and treatments
D. Generating large amounts of oxygen

Answer: D. Generating large amounts of oxygen

Reading Passage:

The growth of urban agriculture has been gaining momentum in recent years, as more people recognize the benefits of growing their own food within cities. Urban gardens not only provide fresh, locally sourced produce, but they also offer a range of environmental, social, and health advantages. Green spaces created by urban agriculture can help absorb carbon dioxide, reduce urban heat island effects, and improve air quality. Furthermore, community gardens can foster social connections, encouraging collaboration and a sense of belonging among residents. Lastly, access to fresh produce can promote healthier eating habits and increased physical activity, contributing to overall well-being.

Question 1:
What are some environmental benefits of urban agriculture?

A. Increased industrialization
B. Lowered property values
C. Carbon dioxide absorption and reduced urban heat island effects
D. Promotion of automobile use

Answer: C. Carbon dioxide absorption and reduced urban heat island effects

Question 2:
How can urban agriculture contribute to social connections within communities?

A. By isolating individuals from one another
B. By discouraging collaboration between residents
C. By fostering a sense of competition among gardeners
D. By encouraging collaboration and a sense of belonging

Answer: D. By encouraging collaboration and a sense of belonging

Question 3:
In what way does urban agriculture promote healthier lifestyles?

A. By reducing access to fresh produce
B. By promoting sedentary habits
C. By increasing access to fast food restaurants
D. By providing access to fresh produce and increasing physical activity

Answer: D. By providing access to fresh produce and increasing physical activity

Reading Passage:

Ocean currents are the continuous and directed movement of seawater caused by various forces acting upon the water, including temperature differences, salinity, and wind. These currents play a crucial role in regulating Earth's climate, as they help to distribute heat from the equator to the poles. Ocean currents are divided into two types: surface currents and deep currents. Surface currents, which make up about 10% of ocean water, are primarily driven by wind and the Earth's rotation. In contrast, deep currents, which account for the remaining 90% of ocean water, are influenced by differences in water density caused by variations in temperature and salinity.

Question 1:
What factors contribute to the formation of ocean currents?

A. Gravity and electromagnetic forces
B. Temperature differences, salinity, and wind
C. Lunar cycles and solar flares
D. Tides and human activities

Answer: B. Temperature differences, salinity, and wind

Question 2:
What role do ocean currents play in regulating Earth's climate?

A. They have no impact on the climate
B. They help distribute heat from the poles to the equator
C. They help distribute heat from the equator to the poles
D. They increase global temperatures

Answer: C. They help distribute heat from the equator to the poles

Question 3:
What is the primary difference between surface currents and deep currents?

A. Surface currents are faster than deep currents
B. Deep currents are influenced by wind, while surface currents are influenced by temperature and salinity
C. Surface currents make up a larger percentage of ocean water than deep currents
D. Surface currents are primarily driven by wind and Earth's rotation, while deep currents are influenced by differences in water density

Answer: D. Surface currents are primarily driven by wind and Earth's rotation, while deep currents are influenced by differences in water density

Reading Passage:

Pollinators play a vital role in our ecosystems by transferring pollen between flowers, enabling plants to reproduce and produce fruit. Bees, butterflies, hummingbirds, and even bats are some of the main pollinators, responsible for the pollination of nearly 75% of the world's most important food crops. However, pollinator populations have been declining due to habitat loss, pesticides, and climate change. This decline has raised concerns about the sustainability of our food systems and the overall health of ecosystems. Conservation efforts to protect pollinators, such as creating pollinator-friendly habitats and reducing pesticide use, are essential to maintaining biodiversity and ensuring food security.

Question 1:
What is the primary role of pollinators in ecosystems?

A. Predation and population control
B. Decomposition of organic matter
C. Transferring pollen between flowers, enabling plant reproduction and fruit production
D. Providing a source of food for higher-level consumers

Answer: C. Transferring pollen between flowers, enabling plant reproduction and fruit production

Question 2:
What is the main cause of pollinator population decline?

A. Human overpopulation
B. Habitat loss, pesticides, and climate change
C. Introduction of invasive species
D. Lack of genetic diversity

Answer: B. Habitat loss, pesticides, and climate change

Question 3:
Why is it important to protect pollinators?

A. To maintain tourism industries
B. To prevent the spread of disease
C. To maintain biodiversity and ensure food security
D. To control pest populations

Answer: C. To maintain biodiversity and ensure food security

Reading Passage:

The ocean is a vast, largely unexplored environment teeming with life and resources. One particularly intriguing aspect of the ocean is the existence of deep-sea hydrothermal vents. These underwater hot springs occur where tectonic plates are diverging, allowing magma to heat the surrounding seawater. The heated water rises from the vents, carrying with it valuable minerals, such as copper, gold, and zinc. Remarkably, these extreme environments are home to unique ecosystems, including species that have never been seen elsewhere. Many of these creatures, such as tube worms and certain bacteria, have adapted to thrive in the absence of sunlight by relying on a process called chemosynthesis for their energy needs.

Question 1:
Where do deep-sea hydrothermal vents typically occur?

A. In shallow coastal waters
B. Where tectonic plates are diverging
C. At the tops of underwater mountains
D. Near river deltas

Answer: B. Where tectonic plates are diverging

Question 2:
Which of the following minerals can be found in the heated water that rises from hydrothermal vents?

A. Copper, gold, and zinc
B. Iron, silver, and magnesium
C. Calcium, potassium, and sulfur
D. Sodium, chloride, and bromine

Answer: A. Copper, gold, and zinc

Question 3:
How do organisms living near hydrothermal vents obtain energy in the absence of sunlight?

A. Through photosynthesis
B. By consuming other organisms
C. Through chemosynthesis
D. By absorbing heat from the vents

Answer: C. Through chemosynthesis

Find the next number in the series.
Question 1:
2 | 6 | 12 | 20 | ?

Answer: A. 30
Explanation: The series represents the consecutive triangular numbers multiplied by 2: (1*2), (3*2), (6*2), (10*2), (15*2).

Question 2:
3 | 9 | 27 | 81 | ?

Answer: C. 243
Explanation: The series shows a geometric progression where each number is multiplied by 3: (3^1), (3^2), (3^3), (3^4), (3^5).

Question 3:
1 | 4 | 9 | 16 | ?

Answer: B. 25
Explanation: The series represents consecutive square numbers: (1^2), (2^2), (3^2), (4^2), (5^2).

Question 4:
1 | 8 | 27 | 64 | ?

Answer: B. 125
Explanation: The series represents consecutive cube numbers: (1^3), (2^3), (3^3), (4^3), (5^3).

Question 5:
2 | 5 | 11 | 23 | ?

Answer: B. 47
Explanation: The series represents consecutive prime numbers multiplied by 2 minus 1: (2·1)-1, (2·2)-1, (2·3)-1, (2·5)-1, (2*7)-1.

Look at this series: 7 | 12 | 19 | 28 | ?
A. 37
B. 38
C. 39
D. 40

Answer: C. 39
Explanation: The series increases by consecutive odd numbers: 5, 7, 9, and so on. The next odd number in the sequence is 11, so 28 + 11 = 39.

Observe this pattern: 1 | 1 | 2 | 3 | 5 | ?
A. 6
B. 7
C. 8
D. 9

Answer: B. 7
Explanation: This series represents the Fibonacci sequence where each number is the sum of the two preceding ones. The next number is 3 + 5 = 7.

Check out this sequence: 1 | 4 | 10 | 22 | ?
A. 43
B. 46
C. 48
D. 52

Answer: B. 46
Explanation: The series represents the following pattern: 1x2, 2x2, 5x2, 11x2. We can see that the multiplier increases by one prime number each time, so the next multiplier should be 23. Thus, the next number is 23x2 = 46.

Consider this series: 1 | 8 | 15 | 22 | ?
A. 28
B. 29
C. 30
D. 36

Answer: D. 36
Explanation: The series increases by a constant difference of 7: 1+7=8, 8+7=15, 15+7=22. To find the next number, add 7 to 22, which equals 36.

Look at this pattern: 2 | 12 | 30 | 56 | ?
A. 80
B. 90
C. 100
D. 110

Answer: C. 100
Explanation: The series represents consecutive triangular numbers multiplied by 2: 1x(1+1)x2, 2x(2+1)x2, 3x(3+1)x2, 4x(4+1)x2. The next number should be 5x(5+1)x2 = 5x6x2 = 100.

Sure, here are five quadratic equation practice questions without numbering, along with the answers and explanations:

Which of the following equations has two distinct real roots?
A. $x^2 - 4x + 4 = 0$
B. $x^2 + 2x + 1 = 0$
C. $x^2 - 6x + 9 = 0$
D. $x^2 - 3x + 2 = 0$

Answer: D. $x^2 - 3x + 2 = 0$
Explanation: For a quadratic equation to have two distinct real roots, its discriminant ($b^2 - 4ac$) must be greater than 0. Only the equation in option D satisfies this condition.

Find the vertex of the following quadratic function: f(x) = 2x^2 - 8x + 3
A. (2, -5)
B. (4, -9)
C. (2, 3)
D. (4, 3)

Answer: A. (2, -5)
Explanation: The vertex of a quadratic function in the form f(x) = ax^2 + bx + c can be found using the formula (h, k), where h = -b/(2a) and k = f(h). Applying this formula, we find the vertex to be (2, -5).

Which of the following represents the quadratic function in vertex form?
A. y = (x - 3)^2 + 5
B. y = (x + 3)^2 - 5
C. y = (x - 5)^2 + 3
D. y = (x + 5)^2 - 3

Answer: A. y = (x - 3)^2 + 5
Explanation: A quadratic function in vertex form is represented by y = a(x - h)^2 + k, where (h, k) is the vertex. Option A represents the quadratic function in vertex form.

Question:
What is the axis of symmetry for the quadratic equation x^2 - 4x + 3 = 0?
A. x = 1
B. x = 2
C. x = 3
D. x = 4

Answer: B. x = 2
Explanation: The axis of symmetry for a quadratic equation in the form ax^2 + bx + c = 0 is given by x = -b/(2a). In this case, a = 1 and b = -4, so the axis of symmetry is x = 2.

Given the quadratic function f(x) = x^2 - 6x + 8, what is the range when x is between 2 and 5?
A. [0, 5)
B. (0, 5]
C. [0, 9)
D. (0, 9]

Answer: B. (0, 5]
Explanation: The vertex of the function is at (3, -1), and the parabola opens upwards. When x is between 2 and 5, the minimum value of the function is -1, occurring at the vertex. The maximum value occurs when x = 5, which is 5^2 - 6(5) + 8 = 5. Thus, the range is (0, 5].

Which expression represents a polynomial?
A. 3x^2 + 2/x
B. 4x^3 - 5x^2 + x + 7
C. 6x^2 - 3√x + 2
D. 2x^4 - x^2 + 1/x^3

Answer: B. 4x^3 - 5x^2 + x + 7
Explanation: A polynomial is an expression composed of variables, coefficients, and constants, combined using addition, subtraction, and multiplication. In option B, the expression consists of terms with whole number exponents, making it a polynomial.

What is the degree of the polynomial 5x^4 - 3x^2 + 7x - 1?
A. 1
B. 2
C. 3
D. 4

Answer: D. 4
Explanation: The degree of a polynomial is the highest exponent of its terms. In this case, the highest exponent is 4, making the degree of the polynomial 4.

Which polynomial is a perfect square trinomial?
A. $x^2 - 4x + 4$
B. $x^2 + 4x + 4$
C. $x^2 - 6x + 9$
D. $x^2 + 6x + 9$

Answer: A. $x^2 - 4x + 4$
Explanation: A perfect square trinomial can be factored into the square of a binomial. In this case, $x^2 - 4x + 4$ can be factored into $(x - 2)^2$.

Which of the following is the simplified form of the expression $(2x^2 - 5x + 3) + (x^2 + 4x - 2)$?
A. $3x^2 - x + 1$
B. $3x^2 - x + 5$
C. $x^2 + 9x + 1$
D. $x^2 - 9x + 1$

Answer: A. $3x^2 - x + 1$
Explanation: To simplify the expression, combine like terms: $(2x^2 + x^2) + (-5x + 4x) + (3 - 2) = 3x^2 - x + 1$.

What is the remainder when the polynomial $3x^3 + 4x^2 - 2x + 1$ is divided by $x - 2$?
A. 23
B. 17
C. 9
D. 1

Answer: B. 17
Explanation: Using synthetic division or the remainder theorem, the remainder when the polynomial is divided by $x - 2$ can be found by evaluating the polynomial at $x = 2$. So, $3(2)^3 + 4(2)^2 - 2(2) + 1 = 24 + 16 - 4 + 1 = 17$.

Which method is most suitable for solving the system of linear equations: x - 3y = 4 and 4x - 12y = 16?
A. Substitution
B. Elimination
C. Graphing
D. Matrix

Answer: B. Elimination
Explanation: The elimination method is most suitable for this system of linear equations because the coefficients of the y variable are multiples of each other, making it easy to eliminate y by adding or subtracting the equations.

What is the solution to the system of linear equations: 2x - y = 3 and x + y = 1?
A. (1, -1)
B. (2, 1)
C. (1, 2)
D. (2, -1)

Answer: C. (1, 2)
Explanation: You can use the elimination method to solve this system. Add the two equations: (2x - y) + (x + y) = 3x = 3 + 1, so x = 1. Now, substitute x = 1 into the second equation: 1 + y = 1, which gives y = 2. The solution is (1, 2).

The system of linear equations 3x - 2y = 6 and 9x - 6y = 18 has:
A. No solution
B. One solution
C. Infinitely many solutions
D. Two solutions

Answer: C. Infinitely many solutions
Explanation: Divide the second equation by 3: 3x - 2y = 6. This equation is the same as the first equation, so the system has infinitely many solutions, as both equations represent the same line.

Which of the following systems of linear equations has no solution?
A. x - y = 2 and x + y = 4
B. 2x + 3y = 5 and 4x + 6y = 10
C. 3x - 4y = 8 and 6x - 8y = 16
D. x + 2y = 6 and x - y = 1

Answer: B. 2x + 3y = 5 and 4x + 6y = 10
Explanation: If you divide the second equation by 2, you get 2x + 3y = 5, which is the same as the first equation. However, the constants on the right side do not match, so the two lines are parallel and do not intersect, resulting in no solution.

The system of linear equations x + y = 3 and 3x - y = 5 can be solved by:
A. Substitution only
B. Elimination only
C. Either substitution or elimination
D. Neither substitution nor elimination

Answer: C. Either substitution or elimination
Explanation: Since the coefficients of y in the two equations have the same absolute value but different signs, the elimination method can be used by adding the two equations. Alternatively, you can solve for y in the first equation (y = 3 - x) and substitute it into the second equation to use the substitution method. Both methods will yield the correct solution.

What is the value of x + y if the system of linear equations 2x + y = 5 and x - 3y = 7 is solved?
A. 4
B. 6
C. 8
D. 10

Answer: B. 6
Explanation: You can use the elimination method to solve this system. Multiply the second equation by 2: 2x - 6y = 14. Subtract the first equation from this new equation: -5y = 9. This gives y = -9/5. Now substitute y = -9/5 into the first equation: 2x - 9/5 = 5, which gives x = 19/5. Adding x and y, we get 19/5 - 9/5 = 10/5 = 6.

Given the system of linear equations 4x - 2y = 6 and 2x + y = 1, what is the value of x?
A. -1
B. 0
C. 1
D. 2

Answer: C. 1
Explanation: Use the elimination method by multiplying the second equation by 2: 4x + 2y = 2. Add the first equation to this new equation: 8x = 8, which gives x = 1.

The system of linear equations x + 2y = 6 and 2x - y = 4 can be solved by:
A. Substitution only
B. Elimination only
C. Either substitution or elimination
D. Neither substitution nor elimination

Answer: C. Either substitution or elimination
Explanation: You can use the substitution method by solving for x in the first equation: x = 6 - 2y, and then substitute it into the second equation. Alternatively, you can use the elimination method by multiplying the first equation by 2: 2x + 4y = 12, and then subtract the second equation from the new equation. Both methods will yield the correct solution.

Which of the following systems of linear equations has infinitely many solutions?
A. x - y = 2 and x + y = 4
B. 2x + y = 6 and x - 2y = 2
C. 3x + 2y = 12 and 6x + 4y = 24
D. x + 2y = 6 and 3x - y = 5

Answer: C. 3x + 2y = 12 and 6x + 4y = 24
Explanation: The second equation is a multiple of the first equation (6x + 4y = 2*(3x + 2y)), which means that both equations represent the same line, and the system has infinitely many solutions.

If 4x + 5y = 20 and 3x - 2y = 4, what is the value of y?
A. 2
B. 3
C. 4
D. 5

Answer: A. 2
Explanation: Multiply the first equation by 3 and the second equation by 4 to make the coefficients of x the same: 12x + 15y = 60 and 12x - 8y = 16. Subtract the second equation from the first equation: 23y = 44, which gives y = 44/23 ≈ 2.

A triangle has angles A, B, and C, and angle A is twice the size of angle B. If angle C measures 40 degrees, what is the size of angle A?
A. 80 degrees
B. 100 degrees
C. 120 degrees
D. 140 degrees

Answer: B. 100 degrees
Explanation: Since the sum of the angles in a triangle equals 180 degrees, and angle A is twice the size of angle B, we can write: A + B + 40 = 180 and A = 2B. Substituting A in the first equation, we get 2B + B + 40 = 180, which simplifies to 3B = 140, giving B = 140/3 ≈ 46.67. Therefore, angle A is twice that size, which is approximately 100 degrees.

A right triangle has legs of length 5 and 12. What is the length of the hypotenuse?
A. 10
B. 13
C. 15
D. 17

Answer: B. 13
Explanation: Use the Pythagorean theorem: $a^2 + b^2 = c^2$, where a and b are the legs and c is the hypotenuse. In this case, $5^2 + 12^2 = c^2$, which simplifies to 25 + 144 = c^2, giving c^2 = 169. The length of the hypotenuse is therefore c = √169 = 13.

The circumference of a circle is 24π. What is the diameter of the circle?
A. 6
B. 8
C. 12
D. 24

Answer: C. 12
Explanation: The formula for the circumference of a circle is C = πd, where d is the diameter. Given the circumference is 24π, we can find the diameter by solving the equation: 24π = πd, which gives d = 24.

A rectangular room has a length of 12 meters and a width of 8 meters. What is the area of the room?
A. 72 square meters
B. 84 square meters
C. 96 square meters
D. 108 square meters

Answer: C. 96 square meters
Explanation: To find the area of a rectangle, multiply its length by its width. In this case, the area is 12 meters * 8 meters = 96 square meters.

A parallelogram has a base of 15 cm and a height of 8 cm. What is the area of the parallelogram?
A. 90 cm^2
B. 100 cm^2
C. 120 cm^2
D. 140 cm^2

Answer: C. 120 cm^2
Explanation: The area of a parallelogram is given by the formula A = base * height. In this case, the area is 15 cm * 8 cm = 120 cm^2.

Which of the following shapes has four equal sides and four equal angles?
A. Rectangle
B. Rhombus
C. Square
D. Trapezoid

Answer: C. Square
Explanation: A square has four equal sides and four equal angles, each measuring 90 degrees.

In a right triangle, which term describes the side opposite the right angle?
A. Hypotenuse
B. Adjacent
C. Opposite
D. Base

Answer: A. Hypotenuse
Explanation: The hypotenuse is the side opposite the right angle in a right triangle. It is also the longest side of a right triangle.

What is the sum of the interior angles of a pentagon?
A. 360 degrees
B. 540 degrees
C. 720 degrees
D. 900 degrees

Answer: B. 540 degrees
Explanation: The sum of the interior angles of any polygon can be calculated using the formula (n-2) × 180, where n is the number of sides. For a pentagon, n = 5, so the sum of the interior angles is (5-2) × 180 = 3 × 180 = 540 degrees.

Which term describes a straight angle?
A. Acute angle
B. Obtuse angle
C. Reflex angle
D. Straight angle

Answer: D. Straight angle
Explanation: A straight angle is an angle that measures exactly 180 degrees, forming a straight line.

Which type of triangle has all sides of equal length?
A. Equilateral triangle
B. Isosceles triangle
C. Scalene triangle
D. Right triangle

Answer: A. Equilateral triangle
Explanation: An equilateral triangle has all three sides of equal length and all three angles equal to 60 degrees.

A rectangular room has a length of 10 feet and a width of 8 feet. What is the area of the room?
A. 40 square feet
B. 60 square feet
C. 80 square feet
D. 100 square feet

Answer: C. 80 square feet
Explanation: The area of a rectangle is found by multiplying its length by its width. In this case, the area is 10 feet × 8 feet = 80 square feet.

A circle has a diameter of 14 inches. What is the approximate circumference of the circle?
A. 22 inches
B. 44 inches
C. 66 inches
D. 88 inches

Answer: B. 44 inches
Explanation: The circumference of a circle is found using the formula C = πd, where C is the circumference and d is the diameter. Using the value of π as approximately 3.14, the circumference is 3.14 × 14 inches ≈ 44 inches.

A cylinder has a radius of 3 inches and a height of 5 inches. What is the approximate volume of the cylinder?
A. 45 cubic inches
B. 90 cubic inches
C. 141 cubic inches
D. 282 cubic inches

Answer: C. 141 cubic inches
Explanation: The volume of a cylinder is found using the formula V = πr²h, where V is the volume, r is the radius, and h is the height. Using the value of π as approximately 3.14, the volume is 3.14 × (3 inches)² × 5 inches ≈ 141 cubic inches.

A square has a side length of 6 inches. What is the perimeter of the square?
A. 12 inches
B. 18 inches
C. 24 inches
D. 36 inches

Answer: C. 24 inches
Explanation: The perimeter of a square is found by multiplying the length of one side by 4. In this case, the perimeter is 6 inches × 4 = 24 inches.

A rectangular prism has a length of 4 inches, a width of 3 inches, and a height of 2 inches. What is the volume of the prism?
A. 6 cubic inches
B. 12 cubic inches
C. 24 cubic inches
D. 48 cubic inches

Answer: C. 24 cubic inches
Explanation: The volume of a rectangular prism is found by multiplying its length, width, and height. In this case, the volume is 4 inches × 3 inches × 2 inches = 24 cubic inches.

Passage 1:
John, an electrician, is considering upgrading his tools. He is weighing the benefits of buying high-quality, expensive tools against the lower cost of more affordable, but less durable options.

Which factor is most important for John when deciding on tools to purchase?
A. The color of the tools
B. The cost and durability of the tools
C. The popularity of the tools
D. The weight of the tools

Answer: B. The cost and durability of the tools
Explanation: The passage states that John is considering the benefits of buying high-quality, expensive tools against the lower cost of more affordable, but less durable options. This implies that cost and durability are his primary concerns.

Passage 2:
In an electrical circuit, a series connection means that the components are connected end-to-end in a single path so that the current flows through each component in turn.

What does a series connection imply about the flow of current?
A. It flows in multiple paths
B. It flows through each component in turn
C. It flows around the components
D. It flows only through some components

Answer: B. It flows through each component in turn
Explanation: The passage states that in a series connection, the current flows through each component in turn, implying that the current follows a single path.

Passage 3:
As a safety precaution, electrical systems should always be grounded. Grounding helps prevent electrical shock and fires by providing a safe path for electricity to follow in the event of a fault or short circuit.

Why is grounding important in electrical systems?
A. To increase the flow of electricity
B. To prevent electrical shock and fires
C. To make the system more efficient
D. To reduce electrical resistance

Answer: B. To prevent electrical shock and fires
Explanation: The passage states that grounding helps prevent electrical shock and fires by providing a safe path for electricity to follow in the event of a fault or short circuit.

Passage 4:
Electricians use a variety of tools, such as wire strippers, pliers, and screwdrivers, to carry out their tasks. These tools help them install, maintain, and repair electrical systems in residential, commercial, and industrial settings.

What is the primary purpose of the tools mentioned in the passage?
A. To design electrical systems
B. To test electrical systems
C. To install, maintain, and repair electrical systems
D. To measure electrical systems

Answer: C. To install, maintain, and repair electrical systems
Explanation: The passage states that electricians use tools like wire strippers, pliers, and screwdrivers to install, maintain, and repair electrical systems in various settings.

Passage 5:
An electrical transformer is a device that can increase or decrease the voltage of an alternating current (AC). By adjusting the voltage, a transformer helps ensure that electrical devices receive the appropriate voltage for safe and efficient operation.

What is the primary function of an electrical transformer?
A. To convert AC to DC
B. To increase or decrease the voltage of an AC
C. To control the flow of current
D. To store electrical energy

Answer: B. To increase or decrease the voltage of an AC
Explanation: The passage states that an electrical transformer's primary function is to increase or decrease the voltage of an alternating current (AC), which helps ensure that electrical devices receive the appropriate voltage.

Which of the following is an active reading technique that helps readers engage with the text and improve their understanding?
A. Skimming through the text quickly
B. Reading without taking notes
C. Highlighting or underlining key points
D. Ignoring headings and subheadings

Answer: C. Highlighting or underlining key points
Explanation: Highlighting or underlining key points while reading helps readers engage with the text and improves their understanding of the material. The other options do not actively involve the reader in the reading process.

Which active reading technique can help readers remember information better and organize their thoughts?
A. Taking breaks while reading
B. Taking notes or summarizing the material
C. Reading multiple sources simultaneously
D. Reading aloud to oneself

Answer: B. Taking notes or summarizing the material
Explanation: Taking notes or summarizing the material as you read helps to reinforce your memory of the information and organize your thoughts on the topic.

When encountering an unfamiliar term while reading, what should you do to improve comprehension?
A. Skip the term and continue reading
B. Look up the term in a dictionary or glossary
C. Replace the term with a familiar word
D. Assume the term is not important

Answer: B. Look up the term in a dictionary or glossary
Explanation: When encountering an unfamiliar term, looking it up in a dictionary or glossary helps improve comprehension by providing the necessary background information.

What is the purpose of previewing a text before reading it in-depth?
A. To avoid reading the entire text
B. To get a general sense of the text's structure and content
C. To identify the author's credentials
D. To find interesting quotes and anecdotes

Answer: B. To get a general sense of the text's structure and content
Explanation: Previewing a text before reading it in-depth allows you to get a general sense of its structure and content, helping you understand the overall context before engaging with the material more closely.

How can asking questions while reading help improve comprehension?
A. By encouraging readers to skip confusing sections
B. By forcing readers to memorize information
C. By engaging readers in a dialogue with the text
D. By distracting readers from the main ideas

Answer: C. By engaging readers in a dialogue with the text
Explanation: Asking questions while reading encourages readers to engage in a dialogue with the text, prompting them to think critically about the material and improving their comprehension.

Passage 1: The blue whale is the largest animal ever to have lived on Earth, even bigger than the dinosaurs. Its heart alone is the size of a small car, and its tongue is so heavy that 50 people could stand on it.

Which of the following statements is true?
A. The blue whale's tongue is the size of a small car.
B. Fifty people can lift the blue whale's heart.
C. Blue whales are larger than any dinosaur.
D. Dinosaurs are larger than blue whales.

Answer: C. Blue whales are larger than any dinosaur.

Passage 2: Recycling helps reduce the amount of waste that goes to landfills and conserves natural resources. By recycling, we are also reducing pollution and saving energy.

What is one benefit of recycling?
A. Generating more waste for landfills
B. Using more natural resources
C. Reducing pollution and saving energy
D. Increasing pollution and energy consumption

Answer: C. Reducing pollution and saving energy

Passage 3: Bees play a crucial role in pollination, which helps plants produce fruits and seeds. Without bees, our food system would face significant challenges, as many crops depend on bees for pollination.

Why are bees important for the food system?
A. They eat harmful insects.
B. They help with pollination, which supports crop growth.
C. They produce honey.
D. They protect plants from diseases.

Answer: B. They help with pollination, which supports crop growth.

Passage 4: Rainforests are home to more than half of the world's plant and animal species. They also absorb carbon dioxide and produce oxygen, making them essential for the planet's health.

What is one reason rainforests are essential for the planet's health?
A. They provide habitat for most plant and animal species.
B. They increase carbon dioxide levels in the atmosphere.
C. They decrease oxygen production.
D. They cause climate change.

Answer: A. They provide habitat for most plant and animal species.

Passage 5: The Great Wall of China was built over 2,000 years ago to protect China from invasions. It stretches over 13,000 miles and is the longest wall in the world.

What was the primary purpose of building the Great Wall of China?
A. To serve as a tourist attraction
B. To provide a transportation route
C. To protect China from invasions
D. To display architectural prowess

Answer: C. To protect China from invasions

Passage: In the small coastal town of Oceanview, a group of dedicated citizens had come together to address a critical issue affecting their community. Over the past decade, the ocean waters had begun to rise, causing the shoreline to erode and threatening homes and businesses located near the coast. The Oceanview Conservation Committee was formed to investigate the cause of this problem and find solutions to protect their town from further damage.

The committee discovered that the primary reason for the rising sea levels was climate change. As global temperatures increased, the polar ice caps began to melt, causing ocean levels to rise and the shoreline to erode. The committee realized that in order to protect their town, they would have to address the issue of climate change and find ways to mitigate its effects on their community.

After extensive research and consultation with experts, the committee decided to focus on three main strategies to combat climate change and protect Oceanview. First, they would work to reduce greenhouse gas emissions in the town by promoting the use of renewable energy sources, such as solar and wind power, and encouraging residents to use energy-efficient appliances and vehicles. Second, they would implement measures to help the community adapt to the effects of climate change by constructing sea walls and reinforcing coastal infrastructure. Finally, the committee would create an education and outreach program to raise awareness about climate change and its impact on the community.

The Oceanview Conservation Committee's efforts were met with support and enthusiasm from the community. As a result, the town saw a significant increase in the use of renewable energy, improved coastal defenses, and a greater understanding of the need to address climate change. Oceanview became a shining example of how a small town could make a big difference in the fight against climate change.

Question 1:
What was the main issue affecting the town of Oceanview?
A. The construction of sea walls
B. The promotion of renewable energy sources
C. The rising ocean waters and eroding shoreline
D. The increase in greenhouse gas emissions

Question 2:
What was the primary cause of the rising sea levels in Oceanview?
A. Coastal development
B. Climate change
C. Overfishing
D. Pollution

Question 3:
Which of the following was NOT one of the strategies the Oceanview Conservation Committee focused on to combat climate change?
A. Reducing greenhouse gas emissions
B. Constructing sea walls
C. Banning fishing in the area
D. Creating an education and outreach program

Question 4:
How did the community of Oceanview respond to the efforts of the Oceanview Conservation Committee?
A. With indifference
B. With opposition
C. With support and enthusiasm
D. With skepticism

Question 5:
What was the overall outcome of the Oceanview Conservation Committee's efforts?
A. The town saw a decline in renewable energy usage.
B. The town became a shining example in the fight against climate change.
C. The town's coastal defenses were weakened.
D. The town experienced an increase in pollution levels.

Passage: Deep within the heart of the Amazon rainforest lies a vast and vibrant ecosystem teeming with life. This incredible region, known as the "lungs of the Earth," plays a crucial role in maintaining the planet's climate by absorbing large amounts of carbon dioxide and producing oxygen. However, in recent years, the Amazon has faced significant threats from deforestation and habitat destruction, which have had far-reaching consequences for the global environment.

Deforestation in the Amazon is primarily driven by the demand for agricultural land and resources. As the world's population continues to grow, the need for food, especially meat, has increased, leading to the expansion of cattle ranching and soybean farming in the rainforest. Additionally, the extraction of valuable resources such as timber and minerals has resulted in the loss of vast areas of pristine forest.

The consequences of deforestation in the Amazon are multifaceted and devastating. As trees are cleared, the carbon dioxide they once absorbed is released back into the atmosphere, contributing to climate change. The loss of habitat also has a profound impact on the incredible biodiversity found in the Amazon, with countless species being pushed to the brink of extinction. Furthermore, the destruction of the rainforest disrupts the lives and livelihoods of the indigenous communities who have depended on the forest for generations.

In response to these threats, conservation organizations and governments have taken steps to protect the Amazon and its unique ecosystem. Efforts have been made to establish protected areas and reserves, promote sustainable land management practices, and support the rights of indigenous communities. While these actions have had some success, much work remains to be done to ensure the long-term survival of the Amazon rainforest and its essential role in the global environment.

Question 1: What role does the Amazon rainforest play in maintaining the planet's climate?
A. It primarily produces carbon dioxide.
B. It absorbs carbon dioxide and produces oxygen.
C. It releases large amounts of methane.
D. It increases global temperatures by absorbing sunlight.

Question 2: Which of the following is the primary driver of deforestation in the Amazon?
A. Tourism
B. Natural disasters
C. Demand for agricultural land and resources
D. Overpopulation of indigenous communities

Question 3: How does deforestation in the Amazon contribute to climate change?
A. By increasing the amount of carbon dioxide absorbed by the forest
B. By releasing carbon dioxide that was once absorbed by the trees
C. By decreasing global temperatures
D. By producing more oxygen in the atmosphere

Question 4: What is one consequence of deforestation in the Amazon for its biodiversity?
A. The number of invasive species decreases.
B. Many species are pushed to the brink of extinction.
C. New habitats are created for endangered species.
D. The overall number of species in the rainforest increases.

Question 5: Which of the following actions has been taken in response to the threats facing the Amazon rainforest?
A. Encouraging the expansion of cattle ranching
B. Establishing protected areas and reserves
C. Removing legal protections for indigenous communities
D. Promoting unsustainable land management practices

Answers:
B. It absorbs carbon dioxide and produces oxygen.
C. Demand for agricultural land and resources
B. By releasing carbon dioxide that was once absorbed by the trees
B. Many species are pushed to the brink of extinction.
B. Establishing protected areas and reserves

Passage: The invention of the printing press in the 15th century revolutionized the way information was disseminated and consumed. Prior to this groundbreaking innovation, books were painstakingly handwritten and copied by scribes, making them scarce and expensive. The printing press, developed by Johannes Gutenberg, made it possible to produce large quantities of books efficiently and cost-effectively, leading to a significant increase in literacy and the spread of ideas throughout the world.

Gutenberg's printing press used movable type, which allowed individual letters and characters to be rearranged and reused for different texts. This system made it much faster to create multiple copies of a single work, as the types could be quickly rearranged to produce new pages. The printing press also facilitated the standardization of language and spelling, as well as the development of new fonts and styles, which made reading more accessible to a wider audience.

The impact of the printing press on society was immense. As books became more affordable and available, people from various social classes could access information and education previously reserved for the elite. The increased availability of books spurred the growth of libraries, which became important centers for learning and knowledge exchange. The printing press also played a key role in the dissemination of scientific discoveries, religious texts, and artistic works, contributing to intellectual and cultural advancements during the Renaissance and beyond.

The printing press has had lasting effects on the world, shaping the course of history and influencing countless aspects of modern society. Today, we continue to see its impact in the widespread availability of printed materials, the development of the internet, and the democratization of information. The invention of the printing press is a testament to the power of human ingenuity and the potential for technology to transform the way we live, learn, and communicate.

Question 1:
What was the primary function of Gutenberg's printing press?
A. To create artistic works
B. To make books more affordable and available
C. To standardize language and spelling
D. To promote the growth of libraries

Question 2:
What was the key innovation of Gutenberg's movable type system?
A. It allowed for faster, more efficient copying of handwritten texts.
B. It eliminated the need for scribes.
C. It made it possible to reuse and rearrange individual letters and characters.
D. It increased the variety of fonts and styles available.

Question 3:
How did the printing press affect literacy rates?
A. It decreased literacy rates due to the scarcity of books.
B. It increased literacy rates by making books more accessible and affordable.
C. It had no effect on literacy rates.
D. It led to a decline in the quality of education.

Question 4:
Which of the following was an important impact of the printing press on society?
A. The decline of libraries
B. The restriction of information to the elite
C. The spread of ideas and increased availability of books
D. The standardization of artistic styles

Question 5:
What lasting effect of the printing press is still seen in modern society?
A. The reliance on handwritten texts
B. The democratization of information
C. The limited availability of printed materials
D. The decline of the internet

Answers:
B. To make books more affordable and available
C. It made it possible to reuse and rearrange individual letters and characters.
B. It increased literacy rates by making books more accessible and affordable.
C. The spread of ideas and increased availability of books
B. The democratization of information

Passage: After a long day at work, Sam returned home and noticed the front door was ajar. Stepping inside, he saw muddy footprints leading from the door to the living room. He quickly checked the living room and discovered that his TV and laptop were missing. The living room window was broken, and there was shattered glass on the floor. Sam immediately called the police to report the incident.

What can be inferred about the intruder's entry?
A. The intruder entered through the front door.
B. The intruder entered through the living room window.
C. The intruder entered through the back door.
D. The intruder entered through the garage.

Answer: B. The intruder entered through the living room window.
Explanation: Since the living room window was broken and there was shattered glass on the floor, it can be inferred that the intruder entered through the window.

What can be inferred about Sam's neighborhood?
A. It is a very safe neighborhood.
B. It is a high-crime neighborhood.
C. It is a gated community.
D. It cannot be determined from the information given.

Answer: D. It cannot be determined from the information given.
Explanation: There is not enough information provided in the passage to make a conclusion about the neighborhood.

Which of the following can be inferred about the intruder?
A. The intruder was specifically targeting Sam.
B. The intruder was looking for valuable items to steal.
C. The intruder was a friend of Sam's.
D. The intruder left a note.

Answer: B. The intruder was looking for valuable items to steal.
Explanation: The passage mentions that the TV and laptop were missing, suggesting that the intruder was stealing valuable items.

What can be inferred about the weather outside?
A. It was raining.
B. It was snowing.
C. It was sunny.
D. It cannot be determined from the information given.

Answer: D. It cannot be determined from the information given.
Explanation: The passage does not provide enough information to determine the weather outside.

What is the most likely reason Sam called the police?
A. To report the muddy footprints.
B. To report the broken window.
C. To report the missing items.
D. To report the open front door.

Answer: C. To report the missing items.
Explanation: Sam called the police to report the incident, which included the missing TV and laptop. While the other options are related to the incident, the primary reason for calling the police would be to report the stolen items.

Passage: The use of electric cars has seen a significant increase in recent years, with more and more people choosing to make the switch from traditional gasoline-powered vehicles. The benefits of electric cars are numerous, including reduced emissions, lower fuel costs, and quieter operation. However, critics argue that the production of electric car batteries is harmful to the environment and that the electricity used to charge them often comes from non-renewable sources. As the debate continues, it is crucial for consumers to weigh the pros and cons of electric cars before making a decision.

What is the author's primary purpose in this passage?
A. To persuade readers to buy an electric car
B. To provide information on the benefits and drawbacks of electric cars
C. To argue against the use of electric cars
D. To describe how electric cars work

Answer: B. To provide information on the benefits and drawbacks of electric cars
Explanation: The author presents both the advantages and disadvantages of electric cars, allowing readers to make an informed decision.

Which of the following best describes the author's point of view?
A. Strongly in favor of electric cars
B. Strongly against electric cars
C. Neutral, presenting both sides of the argument
D. Uncertain about the benefits of electric cars

Answer: C. Neutral, presenting both sides of the argument
Explanation: The author presents both the benefits and drawbacks of electric cars, without showing a strong preference for either side.

What can be inferred about the author's opinion on consumer choice?
A. The author believes that consumers should always choose electric cars.
B. The author believes that consumers should not choose electric cars.
C. The author believes that consumers should make their own informed decisions.
D. The author does not express an opinion on consumer choice.

Answer: C. The author believes that consumers should make their own informed decisions.
Explanation: The author encourages consumers to weigh the pros and cons of electric cars before making a decision, implying that they believe in informed consumer choice.

What is the author's purpose in mentioning the drawbacks of electric cars?
A. To discourage readers from considering electric cars
B. To provide a balanced perspective on the issue
C. To focus on the negative aspects of electric cars
D. To contradict the benefits presented earlier in the passage

Answer: B. To provide a balanced perspective on the issue
Explanation: The author presents both the benefits and drawbacks of electric cars to give readers a comprehensive understanding of the topic.

Based on the passage, which of the following best describes the author's attitude towards the ongoing debate about electric cars?
A. Dismissive of the debate as unimportant
B. Critical of the arguments made by both sides
C. Interested in presenting a fair and balanced view
D. Impatient for a resolution to the debate

Answer: C. Interested in presenting a fair and balanced view
Explanation: The author presents both the benefits and drawbacks of electric cars and encourages consumers to make informed decisions, showing their interest in providing a fair and balanced perspective on the debate.

Passage: The popularity of renewable energy sources has skyrocketed in recent years. Solar energy, in particular, has gained a lot of attention due to its abundance and potential. Solar panels are the primary technology used to harness this energy, and they function by converting sunlight into electricity. There are three primary types of solar panels: monocrystalline, polycrystalline, and thin-film. Each type has its own set of advantages and disadvantages, depending on factors like efficiency, cost, and aesthetics. As the demand for clean energy increases, it is essential for consumers to understand the differences between these solar panel types to make informed decisions.

What is the primary organizational structure of the passage?
A. Chronological order
B. Cause and effect
C. Comparison and contrast
D. Problem and solution

Answer: C. Comparison and contrast
Explanation: The passage focuses on comparing and contrasting the three primary types of solar panels.

Which part of the passage introduces the main topic?
A. The popularity of renewable energy sources has skyrocketed in recent years.
B. Solar energy, in particular, has gained a lot of attention due to its abundance and potential.
C. Solar panels are the primary technology used to harness this energy...
D. There are three primary types of solar panels: monocrystalline, polycrystalline, and thin-film.

Answer: B. Solar energy, in particular, has gained a lot of attention due to its abundance and potential.
Explanation: This sentence introduces solar energy as the main topic and focus of the passage.

What is the primary purpose of the passage?
A. To argue that solar energy is the best renewable energy source
B. To describe the process of converting sunlight into electricity
C. To inform readers about the different types of solar panels
D. To persuade readers to invest in solar energy

Answer: C. To inform readers about the different types of solar panels
Explanation: The passage focuses on providing information about the three primary types of solar panels and their advantages and disadvantages.

Which section of the passage discusses the advantages and disadvantages of solar panel types?
A. The popularity of renewable energy sources has skyrocketed in recent years.
B. Solar energy, in particular, has gained a lot of attention due to its abundance and potential.
C. Solar panels are the primary technology used to harness this energy...
D. Each type has its own set of advantages and disadvantages...

Answer: D. Each type has its own set of advantages and disadvantages...
Explanation: This sentence introduces the comparison of the advantages and disadvantages of each solar panel type.

What function does the last sentence of the passage serve?
A. To summarize the main points of the passage
B. To provide a call to action for the reader
C. To emphasize the importance of understanding solar panel types
D. To introduce a new topic for discussion

Answer: C. To emphasize the importance of understanding solar panel types
Explanation: The last sentence highlights the importance of understanding the differences between solar panel types for consumers to make informed decisions.

The technician's meticulous approach to his work ensured that every detail was carefully considered and addressed. In this context, what does "meticulous" mean?
A. Hasty
B. Thorough
C. Careless
D. Ambiguous

Answer: B. Thorough
Explanation: The context of the sentence shows that the technician's approach led to careful consideration and addressing of every detail, which implies that "meticulous" means thorough.

The electrician used a multimeter to diagnose the circuit's erratic behavior. In this context, what does "erratic" mean?
A. Predictable
B. Unreliable
C. Consistent
D. Steady

Answer: B. Unreliable
Explanation: If the electrician needed to diagnose the circuit, it suggests that the circuit's behavior was not consistent or predictable, so "erratic" means unreliable.

Upon reviewing the electrical schematic, the engineer discovered an anomaly in the design. In this context, what does "anomaly" mean?
A. Improvement
B. Irregularity
C. Standard
D. Copy

Answer: B. Irregularity
Explanation: The engineer discovered something in the design that required attention, so "anomaly" in this context means an irregularity or something unusual.

The apprentice was instructed to fastidiously organize the tools and equipment in the workshop. In this context, what does "fastidiously" mean?
A. Casually
B. Hastily
C. Carelessly
D. Carefully

Answer: D. Carefully
Explanation: Organizing tools and equipment implies a level of care and attention to detail, so "fastidiously" in this context means carefully.

The electrician's astute observation of the circuit led to a more efficient solution. In this context, what does "astute" mean?
A. Careless

B. Shrewd
C. Unobservant
D. Impatient

Answer: B. Shrewd
Explanation: The context of the sentence suggests that the electrician's observation was particularly insightful, so "astute" means shrewd or perceptive.

Choose the word that is closest in meaning to "ameliorate."
A. Worsen
B. Understand
C. Improve
D. Complicate

Answer: C. Improve
Explanation: Ameliorate means to make something better or to improve it.

The word "fastidious" can best be described as:
A. Carefree
B. Attentive to detail
C. Slow-moving
D. Uninterested

Answer: B. Attentive to detail
Explanation: Fastidious refers to someone who is very attentive to detail and careful in their work.

Which of these words means "to make something less severe or intense"?
A. Exacerbate
B. Alleviate
C. Precipitate
D. Encapsulate

Answer: B. Alleviate
Explanation: Alleviate means to make something less severe, intense, or painful.

Select the word that has a similar meaning to "taciturn."
A. Talkative
B. Quiet
C. Energetic
D. Boisterous

Answer: B. Quiet
Explanation: Taciturn describes a person who is reserved and says little.

What is the meaning of "nefarious"?
A. Delightful
B. Wicked
C. Confusing
D. Encouraging

Answer: B. Wicked
Explanation: Nefarious means wicked, evil, or morally wrong.

Based on the context in this sentence, what is the meaning of "abate"? "After the storm, the winds began to abate, and the skies cleared up."
A. Increase
B. Shift
C. Decrease
D. Continue

Answer: C. Decrease
Explanation: The context of the sentence suggests that the winds started to die down after the storm, which means the word "abate" means to decrease or lessen.

What does "gregarious" mean in the following sentence? "Mia is a gregarious person who enjoys going to parties and meeting new people."
A. Shy
B. Sociable
C. Rude
D. Fearful

Answer: B. Sociable
Explanation: The context of the sentence indicates that Mia enjoys socializing, so "gregarious" means sociable or outgoing.

Determine the meaning of "querulous" in this sentence: "The querulous child complained all day about the long car ride."
A. Excited
B. Complaining
C. Content
D. Energetic

Answer: B. Complaining
Explanation: The context of the sentence suggests that the child was unhappy and complaining, which means "querulous" means prone to complaining or expressing dissatisfaction.

In the following sentence, what does "efficacious" mean? "The new medicine proved to be efficacious in treating the patient's symptoms."
A. Harmful
B. Ineffective
C. Effective
D. Expensive

Answer: C. Effective
Explanation: The context of the sentence indicates that the medicine was successful in treating the patient's symptoms, so "efficacious" means effective or producing the desired results.

Based on the context, what is the meaning of "lugubrious" in this sentence? "The lugubrious music at the funeral made everyone feel even more sorrowful."
A. Uplifting
B. Mournful
C. Fast-paced
D. Energetic

Answer: B. Mournful
Explanation: The context of the sentence suggests that the music made people feel more sorrowful, so "lugubrious" means mournful or sad.

When practicing reading passages, which strategy can help improve comprehension?
A. Skimming the text quickly without focusing on details
B. Reading only the first and last sentences of each paragraph
C. Reading the passage multiple times
D. Ignoring difficult words or phrases

Answer: C. Reading the passage multiple times
Explanation: Reading the passage multiple times helps reinforce the understanding of the text and allows the reader to better grasp the main ideas and details, leading to improved comprehension.

Which technique is most helpful for understanding a complex passage?
A. Reading the passage aloud
B. Focusing on the punctuation
C. Memorizing the passage
D. Skipping over difficult sections

Answer: A. Reading the passage aloud
Explanation: Reading the passage aloud can help you understand the structure and meaning of complex sentences and identify the relationships between ideas in the text, improving comprehension.

How can annotating a passage help improve reading comprehension?
A. By making the passage look more interesting
B. By forcing the reader to slow down and consider the meaning of the text
C. By creating a distraction from the text
D. By encouraging the reader to skim through the passage

Answer: B. By forcing the reader to slow down and consider the meaning of the text
Explanation: Annotating a passage, such as underlining important phrases, making notes in the margins, or circling key words, forces the reader to engage more actively with the text, leading to better comprehension.

Which strategy can help you better understand the main idea of a passage?
A. Ignoring the introduction and conclusion
B. Focusing only on the details
C. Identifying and summarizing the main idea in your own words
D. Reading the passage as quickly as possible

Answer: C. Identifying and summarizing the main idea in your own words
Explanation: Identifying and summarizing the main idea in your own words forces you to engage with the text and distill the main points, helping to solidify your understanding of the passage.

What can you do to improve your focus while reading a passage?
A. Listen to loud music while reading
B. Read in a noisy environment
C. Take short breaks when you feel your concentration waning
D. Try to multitask while reading

Answer: C. Take short breaks when you feel your concentration waning
Explanation: Taking short breaks when you feel your focus slipping allows your brain to rest and recharge, helping you maintain your concentration and improve your comprehension of the passage.

In a passage about electrical systems, the term "load" refers to:
A. The physical weight of the electrical equipment
B. The amount of electricity a circuit can safely carry
C. The force of gravity on an electrical component
D. The amount of electrical demand on a circuit

Answer: D. The amount of electrical demand on a circuit
Explanation: In electrical systems, "load" refers to the amount of electrical demand placed on a circuit by devices or equipment connected to it.

A passage describing the differences between AC and DC current would likely focus on:
A. The speed of electrons in a wire
B. The direction in which current flows
C. The color of wires used for each type of current
D. The resistance of different materials to electricity

Answer: B. The direction in which current flows
Explanation: The main difference between AC (alternating current) and DC (direct current) is the direction in which the current flows. AC current changes direction periodically, while DC current flows in a single, constant direction.

When reading a passage about electrical safety, which of the following topics would most likely be discussed?
A. The importance of using the correct wire size for a circuit
B. The aesthetics of electrical installations
C. The history of electricity
D. Famous inventors in the electrical industry

Answer: A. The importance of using the correct wire size for a circuit
Explanation: Electrical safety is a crucial aspect of the industry, and using the correct wire size for a circuit is an important safety measure to prevent overheating and potential fires.

In a passage discussing electrical transformers, the primary function of a transformer is to:
A. Convert mechanical energy into electrical energy
B. Store electrical energy for later use
C. Change the voltage of an alternating current
D. Transmit electricity over long distances

Answer: C. Change the voltage of an alternating current
Explanation: The primary function of an electrical transformer is to change the voltage of an alternating current, either stepping it up (increasing the voltage) or stepping it down (decreasing the voltage).

A passage about grounding in electrical systems would most likely emphasize:
A. The importance of connecting electrical systems to the earth
B. The process of installing electrical wiring in a building
C. The creation of an electrical circuit by connecting components
D. The principles of electrical insulation

Answer: A. The importance of connecting electrical systems to the earth
Explanation: Grounding in electrical systems is the process of connecting the electrical system to the earth, providing a safe path for excess electrical current to flow in the event of a fault or short circuit, reducing the risk of electrical shock and damage to equipment.

Ohm's Law states the relationship between voltage (V), current (I), and resistance (R). Which formula represents this relationship?
A. V = I * R
B. V = I / R
C. V = R / I
D. V = I + R

Answer: A. V = I * R
Explanation: Ohm's Law states that the voltage across a resistor is equal to the product of the current flowing through it and its resistance. The formula representing this relationship is V = I * R.

The power (P) in an electrical circuit is related to voltage (V) and current (I) by which formula?
A. P = V + I
B. P = V / I
C. P = I * R
D. P = V * I

Answer: D. P = V * I
Explanation: The power in an electrical circuit is the product of the voltage across the circuit and the current flowing through it. The formula representing this relationship is P = V * I.

If a 60-watt light bulb operates at 120 volts, how much current (I) does the light bulb draw?
A. 0.5 amps
B. 2 amps
C. 60 amps
D. 7200 amps

Answer: A. 0.5 amps
Explanation: Using the power formula P = V * I, we can solve for current by dividing both sides by voltage: I = P / V. Plugging in the values, we get I = 60 watts / 120 volts = 0.5 amps.

An electrical circuit has a resistance of 10 ohms and a current of 2 amps. What is the voltage across the circuit?
A. 5 volts
B. 8 volts
C. 12 volts
D. 20 volts

Answer: D. 20 volts
Explanation: Using Ohm's Law (V = I * R), we can calculate the voltage across the circuit by multiplying the current by the resistance: V = 2 amps * 10 ohms = 20 volts.

A circuit has a power of 100 watts and a current of 4 amps. What is the voltage across the circuit?
A. 25 volts
B. 40 volts
C. 96 volts
D. 400 volts

Answer: B. 40 volts
Explanation: Using the power formula P = V * I, we can solve for voltage by dividing both sides by current: V = P / I. Plugging in the values, we get V = 100 watts / 4 amps = 40 volts.

A journeyman electrician is working on a project that requires 25 feet of conduit per section, and there are 12 sections. How many feet of conduit does he need in total?
A. 275 feet
B. 300 feet
C. 325 feet
D. 350 feet

Answer: C. 325 feet
Explanation: To find the total length of conduit needed, multiply the length per section by the number of sections: 25 feet * 12 sections = 325 feet.

A 500-watt appliance operates for 4 hours per day. How much energy does it consume in one week?
A. 2,000 watt-hours
B. 8,000 watt-hours
C. 14,000 watt-hours
D. 56,000 watt-hours

Answer: D. 56,000 watt-hours
Explanation: First, find the daily energy consumption by multiplying the wattage by the number of hours: 500 watts * 4 hours = 2,000 watt-hours. Then, multiply the daily consumption by 7 days to find the weekly consumption: 2,000 watt-hours * 7 days = 56,000 watt-hours.

If a motor runs at 120 revolutions per minute (RPM), how many revolutions will it complete in 2.5 minutes?
A. 200 revolutions
B. 240 revolutions
C. 300 revolutions
D. 450 revolutions

Answer: D. 450 revolutions
Explanation: To find the total number of revolutions in 2.5 minutes, multiply the RPM by the number of minutes: 120 RPM * 2.5 minutes = 300 revolutions.

An electrical supply store has a 15% discount on a box of wire connectors that regularly cost $80. What is the sale price?
A. $68
B. $72
C. $75
D. $85

Answer: A. $68
Explanation: To find the sale price, first calculate the discount amount: 15% of $80 = 0.15 * $80 = $12. Then, subtract the discount from the original price: $80 - $12 = $68.

A 10-amp circuit breaker protects a circuit with a resistance of 24 ohms. What is the maximum voltage the circuit can handle before the breaker trips?
A. 120 volts
B. 240 volts
C. 480 volts
D. 960 volts

Answer: B. 240 volts
Explanation: Using Ohm's Law (V = I * R), calculate the maximum voltage by multiplying the current (10 amps) by the resistance (24 ohms): 10 amps * 24 ohms = 240 volts.

If 5x - 2 > 13, which of the following represents all possible values of x?
A. x > 3
B. x > 5
C. x > 7
D. x < 3

Answer: A. x > 3
Explanation: First, add 2 to both sides of the inequality: 5x > 15. Then, divide by 5: x > 3.

If |4x - 9| ≤ 15, which of the following represents all possible values of x?
A. -1.5 ≤ x ≤ 6
B. 1.5 ≤ x ≤ 6
C. -6 ≤ x ≤ 1.5
D. -1.5 ≤ x ≤ 9

Answer: A. -1.5 ≤ x ≤ 6
Explanation: The inequality can be split into two separate inequalities: 4x - 9 ≤ 15 and -(4x - 9) ≤ 15. Solve each inequality separately:
4x - 9 ≤ 15:
Add 9: 4x ≤ 24
Divide by 4: x ≤ 6

-(4x - 9) ≤ 15:
Distribute the negative sign: -4x + 9 ≤ 15
Subtract 9: -4x ≤ 6
Divide by -4 and reverse the inequality sign: x ≥ -1.5

Combining the results: -1.5 ≤ x ≤ 6.

If -3x + 4 < 13, which of the following represents all possible values of x?
A. x < -3
B. x < 3
C. x > -3
D. x > 3

Answer: D. x > 3
Explanation: First, subtract 4 from both sides: -3x < 9. Then, divide by -3 and reverse the inequality sign: x > 3.

If 2x - 5 ≥ 9, which of the following represents all possible values of x?
A. x ≥ 2
B. x ≥ 7
C. x ≥ 9
D. x ≥ 14

Answer: B. x ≥ 7
Explanation: First, add 5 to both sides of the inequality: 2x ≥ 14. Then, divide by 2: x ≥ 7.

If |x - 3| > 10, which of the following represents all possible values of x?
A. x < -7 or x > 13
B. x < -10 or x > 13
C. x < -7 or x > 10
D. x < -13 or x > 7

Answer: A. x < -7 or x > 13
Explanation: The inequality can be split into two separate inequalities: x - 3 > 10 and -(x - 3) > 10. Solve each inequality separately:
x - 3 > 10:
Add 3: x > 13
-(x - 3) > 10:
Distribute the negative sign: -x + 3 > 10
Subtract 3: -x > 7
Divide by -1 and reverse the inequality sign: x < -7

Combining the results: x < -7 or x > 13.

PASSAGE: In the electrical industry, grounding is a crucial safety measure that protects people and equipment from electrical faults. Grounding creates a low-resistance path to the earth, which allows the safe dissipation of electrical energy when a fault occurs. A well-grounded system is essential to prevent hazardous voltage levels from building up during a fault, which can lead to electrocution, equipment damage, or fires. Grounding also helps stabilize the voltage levels in a system, reducing the likelihood of electrical disturbances like voltage spikes and surges. Electrical codes and regulations have strict requirements for grounding practices to ensure the safety of both workers and the public.

What is the main purpose of grounding in the electrical industry?
A. To increase the resistance of electrical systems
B. To protect people and equipment from electrical faults
C. To improve energy efficiency
D. To prevent voltage spikes and surges

Answer: B. To protect people and equipment from electrical faults
Explanation: The passage explains that grounding is a crucial safety measure that protects people and equipment from electrical faults by creating a low-resistance path to the earth.

Which of the following is a benefit of a well-grounded system mentioned in the passage?
A. Increased energy efficiency
B. Prevention of hazardous voltage levels during a fault
C. Enhanced system performance
D. Improved power quality

Answer: B. Prevention of hazardous voltage levels during a fault
Explanation: The passage states that a well-grounded system prevents hazardous voltage levels from building up during a fault, reducing the risk of electrocution, equipment damage, or fires.

What does grounding help stabilize, according to the passage?
A. Energy consumption
B. Voltage levels
C. Current flow
D. Electrical resistance

Answer: B. Voltage levels
Explanation: The passage states that grounding helps stabilize the voltage levels in a system, reducing the likelihood of electrical disturbances.

Which of the following is NOT mentioned as a consequence of inadequate grounding?
A. Equipment damage
B. Increased energy consumption
C. Electrocution
D. Fires

Answer: B. Increased energy consumption
Explanation: The passage does not mention increased energy consumption as a consequence of inadequate grounding. It mentions electrocution, equipment damage, and fires as possible outcomes.

What do electrical codes and regulations focus on regarding grounding practices?
A. Energy efficiency
B. Worker and public safety
C. System performance
D. Cost-effectiveness

Answer: B. Worker and public safety
Explanation: The passage states that electrical codes and regulations have strict requirements for grounding practices to ensure the safety of both workers and the public.

PASSAGE: The fascinating world of bees is much more complex than many people realize. These small yet essential insects play a critical role in the pollination of plants, which ultimately contributes to the production of fruits, vegetables, and other plant-based foods. One of the most well-known species, the honeybee, is also responsible for producing honey, a natural sweetener used in various culinary applications.

Honeybees live in highly organized colonies, each containing a single queen, thousands of female worker bees, and hundreds of male drones. The queen bee's primary role is to lay eggs and maintain the population of the colony. Worker bees, on the other hand, are responsible for foraging for nectar and pollen, tending to the young, and producing honey. Drones, meanwhile, exist solely for the purpose of mating with the queen.

Communication within a honeybee colony is intricate and relies on a series of complex behaviors and chemical signals. One such behavior is the "waggle dance," which worker bees use to inform their fellow bees about the location of a nectar source. By performing this dance, they can provide precise directions to the source, enabling other bees to find it with remarkable accuracy.

Despite their small size, bees face a variety of challenges, including habitat loss, pesticide exposure, and climate change. These factors have contributed to the decline of bee populations in recent years, raising concerns about the potential impacts on agriculture and food production. As a result, many people are taking steps to support and protect bees by creating bee-friendly habitats, reducing pesticide use, and advocating for policies that promote their conservation.

What is the primary role of the queen bee in a honeybee colony?
A. To produce honey
B. To forage for nectar and pollen
C. To lay eggs and maintain the colony's population
D. To perform the waggle dance

Answer: C. To lay eggs and maintain the colony's population
Explanation: The passage states that the queen bee's primary role is to lay eggs and maintain the population of the colony.

Which of the following is NOT a responsibility of worker bees?
A. Tending to the young
B. Mating with the queen
C. Foraging for nectar and pollen
D. Producing honey

Answer: B. Mating with the queen
Explanation: The passage explains that worker bees are responsible for foraging for nectar and pollen, tending to the young, and producing honey, but it is the drones that mate with the queen.

What is the purpose of the "waggle dance" performed by worker bees?
A. To attract a mate
B. To communicate the location of a nectar source
C. To establish dominance within the colony
D. To signal the presence of danger

Answer: B. To communicate the location of a nectar source
Explanation: The passage states that the waggle dance is a behavior worker bees use to inform their fellow bees about the location of a nectar source.

Which of the following factors is NOT mentioned as contributing to the decline of bee populations?
A. Habitat loss
B. Pesticide exposure
C. Climate change
D. Overpopulation

Answer: D. Overpopulation
Explanation: The passage does not mention overpopulation as a factor contributing to the decline of bee populations. It discusses habitat loss, pesticide exposure, and climate change as contributing factors.

How are people supporting and protecting bees, according to the passage?
A. By creating bee-friendly habitats
B. By using more pesticides
C. By promoting beekeeping as a hobby
D. By importing more bees from other countries

Answer: A. By creating bee-friendly habitats

The exploration of space has been a captivating and inspiring journey, marked by numerous milestones and discoveries. From the first artificial satellite, Sputnik 1, launched by the Soviet Union in 1957, to the Apollo 11 mission that landed humans on the moon in 1969, these achievements have transformed our understanding of the universe and our place within it.

One of the most intriguing aspects of space exploration is the search for extraterrestrial life. Scientists have developed various methods to detect the presence of life on other planets, such as searching for biosignatures in the atmosphere or examining the surface for evidence of water. The discovery of exoplanets, or planets orbiting stars outside our solar system, has further fueled the search for life beyond Earth. To date, thousands of exoplanets have been identified, with some residing in the habitable zones of their stars, where conditions could potentially support life.

Robotic missions, such as the Mars rovers, have played a crucial role in exploring the potential for life on other planets within our solar system. These missions have provided valuable data on the geological and atmospheric conditions of these celestial bodies, helping scientists to better understand their potential to host life.
In addition to the search for life, space exploration has yielded numerous technological advancements. The development of the Hubble Space Telescope, for example, has revolutionized our understanding of the cosmos by capturing detailed images of distant galaxies and celestial phenomena. Similarly, the International Space Station (ISS) has allowed astronauts to conduct experiments and research in the unique environment of microgravity, leading to breakthroughs in various fields, from biology to materials science.
As we continue to push the boundaries of space exploration, new challenges and opportunities arise. The future of space travel may involve the establishment of permanent human settlements on other planets, such as Mars, and the development of advanced propulsion systems that enable faster and more efficient travel through the cosmos.

Which milestone marked the first time humans landed on the moon?
A. Sputnik 1
B. Apollo 11
C. Hubble Space Telescope
D. International Space Station

Answer: B. Apollo 11

What is the primary goal of searching for biosignatures in the atmosphere of other planets?
A. To study the planet's geology
B. To detect the presence of life
C. To determine the planet's distance from its star
D. To understand the planet's weather patterns

Answer: B. To detect the presence of life

What has the discovery of exoplanets contributed to the search for life beyond Earth?
A. It has provided direct evidence of extraterrestrial life
B. It has identified planets in habitable zones with potential to support life
C. It has allowed for the development of new space telescopes
D. It has led to the establishment of permanent human settlements on other planets

Answer: B. It has identified planets in habitable zones with potential to support life

Which of the following technological advancements has revolutionized our understanding of the cosmos by capturing detailed images of distant celestial objects?
A. Sputnik 1
B. Apollo 11
C. Hubble Space Telescope
D. International Space Station

Answer: C. Hubble Space Telescope

What is one possible future development in the field of space exploration, as mentioned in the passage?
A. The discovery of intelligent extraterrestrial life
B. The establishment of permanent human settlements on other planets
C. The complete mapping of our galaxy
D. The development of time travel

Answer: B. The establishment of permanent human settlements on other planets

In recent years, the rise of electric vehicles (EVs) has transformed the automotive industry, offering consumers an eco-friendly alternative to traditional gasoline-powered cars. These vehicles, powered by electric motors and rechargeable battery packs, are gaining popularity due to their reduced environmental impact, lower operating costs, and advances in battery technology.

One significant benefit of electric vehicles is their potential to reduce greenhouse gas emissions. Traditional internal combustion engine (ICE) vehicles emit carbon dioxide and other pollutants, contributing to climate change and poor air quality. In contrast, EVs produce zero tailpipe emissions, and their overall emissions are generally lower, particularly when charged using renewable energy sources like solar or wind power.

Another advantage of EVs is their lower operating costs. Electric vehicles typically require less maintenance than their ICE counterparts, as they have fewer moving parts and do not need oil changes or exhaust system repairs. Additionally, the cost of electricity to charge an EV is often lower than the cost of gasoline on a per-mile basis, resulting in significant long-term savings for the owner.

Battery technology has seen rapid improvements in recent years, with increased energy density and reduced costs making EVs more practical and accessible. This progress has led to longer driving ranges and faster charging times, addressing some of the primary concerns of potential EV buyers. The development of solid-state batteries, which promise even higher energy density and improved safety, is an exciting prospect for the future of electric vehicles.

Despite their numerous benefits, electric vehicles face several challenges. The limited availability of charging infrastructure, particularly in rural areas, can create a barrier to EV adoption. Furthermore, the production and disposal of batteries raise environmental concerns, as mining for raw materials can cause habitat destruction and pollution, while the disposal of used batteries can contribute to waste issues.

As the electric vehicle market continues to grow and evolve, it is essential to address these challenges to ensure the sustainable development of the industry and the broader transition towards cleaner transportation.

What is the primary source of power for electric vehicles?
A. Gasoline
B. Diesel
C. Electric motors and rechargeable battery packs
D. Hydrogen fuel cells

Answer: C. Electric motors and rechargeable battery packs

Which of the following is a benefit of electric vehicles compared to internal combustion engine vehicles?
A. Higher tailpipe emissions
B. Increased maintenance requirements
C. Reduced greenhouse gas emissions
D. Higher operating costs

Answer: C. Reduced greenhouse gas emissions

What recent advancement in battery technology has contributed to the increased practicality and accessibility of electric vehicles?
A. Longer driving ranges and faster charging times
B. Reduced energy density
C. Increased reliance on fossil fuels
D. Decreased battery life

Answer: A. Longer driving ranges and faster charging times

Which of the following challenges is associated with the adoption of electric vehicles?
A. Limited availability of charging infrastructure
B. Increased greenhouse gas emissions
C. Higher per-mile operating costs
D. Increased dependence on oil

Answer: A. Limited availability of charging infrastructure

What environmental concern is associated with the production and disposal of electric vehicle batteries?
A. Increased air quality
B. Decreased greenhouse gas emissions
C. Habitat destruction and pollution from mining raw materials
D. Reduced waste issues

Answer: C. Habitat destruction and pollution from mining raw materials

The rise of urban gardening has become a global phenomenon, with city dwellers around the world turning to creative methods of growing their own produce. In response to the increasing demand for locally sourced, organic, and sustainable food, urban gardening enthusiasts are transforming underutilized spaces into thriving green oases.

One popular approach to urban gardening is vertical gardening, which maximizes the use of limited space by growing plants on vertical surfaces, such as walls or fences. This method not only allows for a higher density of plants in a smaller area but also improves air quality by filtering pollutants and producing oxygen.

Another innovative urban gardening technique is hydroponics, which involves growing plants in nutrient-rich water rather than soil. Hydroponic systems can be highly efficient in terms of water and space usage, as well as providing the gardener with greater control over the growing conditions. This method is particularly well-suited to urban environments where space is at a premium and soil quality may be poor.

Community gardens are another form of urban gardening that fosters social interaction and environmental awareness. These shared spaces provide an opportunity for people to come together, learn from one another, and develop a deeper connection to their local environment. Community gardens can also serve as educational spaces where children and adults can learn about sustainable agriculture, food systems, and the importance of biodiversity.

Urban gardening initiatives have been shown to have numerous benefits, including promoting local food production, reducing food miles, and lowering the environmental impact of food transport. Additionally, urban gardens can contribute to improved mental well-being, as the act of nurturing plants and engaging with nature has been shown to reduce stress and promote relaxation.

As the world's population continues to urbanize, and the demand for sustainable food production grows, urban gardening offers a promising solution for creating healthier, more sustainable cities. By transforming underused spaces into thriving, productive green spaces, urban gardeners are not only improving the quality of life for city dwellers but also working towards a more sustainable future.

What is one of the primary reasons behind the rise of urban gardening?
A. Increased demand for locally sourced, organic, and sustainable food
B. A desire for more green space in rural areas
C. The need for more recreational parks
D. The growth of the ornamental plant industry

Answer: A. Increased demand for locally sourced, organic, and sustainable food

Which urban gardening method focuses on growing plants on vertical surfaces?
A. Hydroponics
B. Community gardening
C. Vertical gardening
D. Container gardening

Answer: C. Vertical gardening

What is a unique characteristic of hydroponic gardening?
A. Growing plants in soil
B. Growing plants in nutrient-rich water
C. Growing plants on vertical surfaces
D. Growing plants in shared community spaces

Answer: B. Growing plants in nutrient-rich water

What is one benefit of community gardens?
A. They require less maintenance than other types of urban gardens
B. They provide a space for social interaction and environmental awareness
C. They require fewer resources, such as water and nutrients
D. They allow for more efficient use of vertical space

Answer: B. They provide a space for social interaction and environmental awareness

What is one way urban gardening can contribute to a more sustainable future?
A. By promoting local food production and reducing food miles
B. By increasing the demand for non-organic food
C. By encouraging the use of more pesticides and fertilizers
D. By focusing on ornamental plants rather than edible ones

Answer: A. By promoting local food production and reducing food miles